Brooklyn Historic Railway Association

Bob Diamond
Founder, President

Greg Castillo,
Vice President, Implementation & Quality Control
Assurance

Brian Kassel
Vice President, Design & Planning

Dylan Cepeda,
BHRA Publication Editor, Primary Researcher,
Graphics Design

Brooklyn Historic Railway Assn.
599 East 7th Street   Ste 5A
Brooklyn, NY 11218
rdiamond@brooklynrail.net

www.brooklynrail.net

Forward

This Document Was Created Circa 1989. While the
Engineering Drawings and Concepts are Still Valid,
Portions of the Narrative May No Longer Be Current

The New York Times

July 21, 1994

A Move to Revive Brooklyn's Trolleys

By DENNIS HEVESI

Bob Diamond has a dream that he hopes will return the dodgers to Brooklyn -- trolley dodgers, that is.

The original name of the borough's semi-sanctified baseball team was the Brooklyn Trolley Dodgers, derived from the fact that at the turn of the century, players and fans had to scoot out of the way of those newfangled, electrified street cars to get to the ball field.

Mr. Diamond, 34, president of the Brooklyn Historic Railway Association, would love to bring back those oldfangled trolleys, complete with clanging bells and overhead cables strung from vintage, filigreed gaslight-like lampposts.

And perhaps not surprisingly, given the imminent return of trolleys to 42d Street in Manhattan, some officials see Mr. Diamond's scheme as only half harebrained. Nostalgia and Other Benefits

"You know what?" said City Councilman Noach Dear, chairman of the Council's Transportation Committee, "He may have started the ball rolling for us to take a very serious look at trolley service in the other boroughs as a viable transportation alternative."

Mr. Diamond envisions a trolley loop around Brooklyn's downtown area, embracing the Middle Eastern community along Atlantic Avenue, the civic center, Fulton Mall, the MetroTech complex, the anchorages of the Manhattan and Brooklyn Bridges, Brooklyn Heights and the Promenade. A branch would run to Grand Army Plaza as a link to Prospect Park, the Brooklyn Botanic Gardens and the Brooklyn Museum.

"Sure it would be nostalgic and a great tourist attraction," said Mr. Diamond, "but there are other benefits." He points out that trolleys, which disappeared in Brooklyn in the 1950's, cause little, if any, pollution, and are much quieter than buses. They also have 40-to-60-year lifespans, whereas buses often break down within a decade.

Last month, the City Council approved a plan to build a trolley line along 42d Street, connecting one side of Manhattan to the other. Officials have selected four consortiums to makes bids for the $135-million project. The companies would lay track, provide cars and operate the system as a franchise. The system could open by 1997. Some Track Remains

Floyd Lapp, director of the City Planning Department's transportation division, said his department had looked at a transportation loop for downtown Brooklyn in the last year. "A bus loop would be less capital-intensive," he said. "Somebody's got to pay for the track and wire and, in some cases, street reconstruction."

Still, Mr. Diamond remains confident, saying his hope is "to approach the companies that don't get the 42d Street contract to see if they are willing to back a trolley line in downtown Brooklyn." He estimates that the five-mile system with a 14-car fleet would cost $50 million.

Mr. Diamond said power poles could be fashioned like antique street lights, with "pinky-thick wires" 17 to 22 feet above the roadway. In the 1950's, he said, "a lot of trolley track in Brooklyn was reconstructed, then paved over," as buses took over. Some of that track, he believes, could be uncovered and reused.

Despite skepticism for the idea in the short term, some city officials believe trolleys may have a future in the city.

They will "never return to their former dominance in Brooklyn," said Gerard Soffian, the city Transportation Department's assistant commissioner for traffic planning. "But we do see a place for light rail in the city's transportation system. It works. It's very efficient and reliable, and people like it."

Pointing out that Mr. Diamond's proposal was working its way through the bureaucratic maze, Mr. Soffian said, "It has a lot going for it."

And, he added, "Nobody should ever count Bob Diamond out." Creator of Trolley Museum

Mr. Diamond, a round-faced fellow with a sheepish smile, is responsible for unearthing the oldest subway tunnel in the world, built in 1844. By poring over old Brooklyn Eagle news articles and municipal blueprints, he pinpointed the tunnel's location beneath Atlantic Avenue, between Furman Street and Boerum Place. With the aid of volunteers from his nonprofit railway association, he dug out the half-mile, arched tunnel.

In recent years, he has created the Brooklyn Trolley Museum on an old pier at 141 Beard Street in Red Hook, where four street cars are in a constant state of refurbishment.

These days, Mr. Diamond spends most of his time laying track and scraping paint and hoisting engines at his museum. Would he like to be a conductor, tipping his hat and clanging the bell, if the city and some company makes his dream come true?

"Nah," he said. "I'm ready for a desk job."

Photo: The Brooklyn Historic Railway Association has started a campaign to bring back the oldfangled trolleys to Brooklyn, and possibly other boroughs. A trolley served the Flatbush Avenue route in the 1930's. (Culver Pictures)

TROLLEY'S NO FOLLY FAN WANTS TO BRING BACK STREETCARS

BY MICHAEL O. ALLEN / NEW YORK DAILY NEWS / Sunday, July 21, 1996, 12:00 AM

ROBERT Diamond's voice mesmerizes his listeners as he dreams aloud about the return of Brooklyn trolleys 40 years after the borough's last streetcar was dumped into an Avenue J quicksand pit. At his telling, the No. 3 Fulton Landing Trolley, built in 1897, comes to life. Its bells clang, gears accelerate noisily, wheels clack as they round a corner and head for the straightaway.

His dream is to bring the vision back to life, restoring a piece of 19th-century Americana to late 20th-century Brooklyn. "When you swing out to the waterfront, you'll see a beautiful view of the New York Harbor, with the water and the Statue of Liberty, Governors Island and the New York City skyline," said Diamond, president of the Brooklyn Historic Railway Association. "Hopefully, it'll be sunny and 76 degrees," he added.

Diamond, 36, has devoted his life to trolleys since his 1980 discovery of the Atlantic Ave. Long Island Rail Road tunnel that had long been given up for lost. He formed the railway association in 1982, and began collecting vintage trolley cars. He's now got four, and is raising funds for a fifth. With about $60,000 in donated funds and equipment, he and some friends hand-built the first 700 feet of what Diamond envisions as a three-quarter-mile heritage trolley line that will connect two piers along the Red Hook waterfront with local bus routes.

They still have 3,200 feet to go. But Diamond hopes to complete them by fall if he can cut through bureaucratic tie-ups that have delayed payment of a $210,000 federal transportation grant awarded for the project in April 1995.

The effort, part of a new trolley museum, is the first phase of an even grander dream that has won tentative backing from some city officials. Diamond eventually hopes to build a downtown Brooklyn trolley loop that would link the Fulton Mall, the MetroTech complex, the Brooklyn and Manhattan bridges, Brooklyn Heights and the Promenade. He sees the trolleys as a tourist attraction recalling the days when Brooklyn boasted a vast streetcar network with about 500 miles of rail lines. The trolley era ended in 1956, when the last cars were dumped into a quicksand pit at Avenue J and E. 92d St., now the site of a high school.

Diamond, who has studied the history of Brooklyn mass transit, said bringing back trolleys would provide a pollution-free and long-lasting transportation alternative. "They took my life over," Diamond said. "It's a love of history and a love of 19th-century technology".

TUNNEL TOUR The Brooklyn History Railway Association will conduct a tour of the Atlantic Ave. Tunnel next Sunday to raise money for the trolley line. Call the association at (718) 941-3160 for information.

October 24, 1999
NEIGHBORHOOD REPORT: RED HOOK

NEIGHBORHOOD REPORT: RED HOOK; A Desire Named Streetcar

By JULIAN E. BARNES

For years, Bob Diamond has pursued his quirky dream of resurrecting Brooklyn's trolleys. Until this month, the furthest he had gotten was a short line outside a Red Hook warehouse on Van Brunt Street.

But next Sunday, Mr. Diamond will throw the switch on a 1,500-foot trolley run along privately owned waterfront land that will double the length of his line. He hopes it will build support for his proposal to extend the tracks onto city streets.

The City Planning Commission has scheduled a Nov. 10 hearing on the proposal, which would put a railway on Brooklyn's streets for the first time since the last trolley line closed in 1956. If the city gives the go-ahead, Mr. Diamond and his partner, Greg Castillo, say, they can finish building a loop running through a half-dozen blocks in south Red Hook by spring.

Much of their material is donated, and Mr. Diamond and Mr. Castillo do most of the work. So far they have spent close to $400,000, part of it from a Federal grant of $209,000.

Mr. Diamond's grandest dream, of building a trolley line to downtown Brooklyn, is still a long way from reality. But thanks to a new innovation, his simpler plans for a tourist line through Red Hook appears more realistic.

Last summer, Mr. Diamond, an electrical engineer by training, developed a special transformer that allows his trolleys to run on regular Con Edison electric lighting power. The original city trolleys relied on power substations, but Mr. Diamond's transformer, about the size of a trash can, can be hung from the poles that suspend trolley wires.

The prototype transformer powers his newly extended line, and should provide enough energy for his proposed south Red Hook loop. So far the progress has impressed Mr. Diamond's landlord and sponsor, Greg O'Connell.

"He's moving along," said Mr. O'Connell, a Red Hook developer. "This trolley will never be the BMT, but you should see the reactions when people ride it."

The trolley line is the main feature of Mr. Diamond's and Mr. Castillo's Brooklyn Trolley Museum at 499 Van Brunt Street. Rides will be offered on Sundays, weather permitting. "I am hoping people will come out and say this is a cool thing," Mr. Diamond said.

As the trolley runs along the Red Hook waterfront, passengers will see some of the city's most vivid waterfront views and get a feel of what it was like in the glory days of Brooklyn's trolleys, Mr. Castillo said.

"We are going to make this like Mr. Rogers's neighborhood, with an old Brooklyn feel," he said. "It will be one of the prettiest train lines anywhere." JULIAN E. BARNES

Photo: An engineer wants to restore Brooklyn's trolleys to their former glory. (Rebecca Cooney for The New York Times) Chart: "TIMELINE: The Artful Dodger" 1854 -- The first horse-powered streetcar line in Brooklyn opens on July 3, with four horses pulling each car. The line runs from Fulton Ferry down Myrtle Avenue to Marcy Street. The fare is 4 cents. By 1890, some 8,000 horses are used to pull streetcars. 1890 -- The Coney Island Avenue line, Brooklyn's first electric trolley, goes into service on April 20, running from Prospect Park down Coney Island Avenue and then to Brighton Beach. The fare is a nickel. 1920 -- At its peak, the Brooklyn Rapid Transit Company, which owns all but three trolley lines in Brooklyn, has 600 miles of track, 5,500 trolley cars, 125 routes, 20 car barns, 6 terminals, a hotel and an amusement park. 1956 -- The last two Brooklyn trolley lines close on Halloween. No. 35 ran from 92nd Street down Church Avenue and then along 37th street and 39th Street to Bush Terminal. No. 50 ran along MacDonald Ave from Church Street to Avenue Z before turning onto Coney Island. A ride cost 15 cents. (Source: Brooklyn Historic Railway Association)

DOWNTOWN BROOKLYN LIGHT RAIL LINK

BROOKLYN CITY RAILROAD PROJECT

860478GFK; 870997GFK

CEQR NO. 87-141

MARCH, 1989

PREPARED BY:

ROBERT DIAMOND
PRESIDENT
BROOKLYN HISTORIC RAILWAY ASSN.
599 EAST 7TH STREET
BROOKLYN, NY 11218

PHONE: 718-941-3160

5/18/84 2:15 Pm 4th Fl Conference Rm.

Gail Benjamin
Mark Landon
Sylvia Perelli } CEQR: DEP, City Planning Comm.
Barbara Murray
Henry Colon - DOT
Robert Diamond - BHRA
　　Gerry Alasstein - DOT

BROOKLYN CITY RAILROAD PROJECT

ENVIRONMENTAL SCOPE OF WORK

TABLE OF CONTENTS

Page Number

PROJECT DESCRIPTION

Background
The Proposed Project
Public Actions

PREPARATION OF THE EIS

Task 1 – Project Scope – Local/Regional

Task 2 – Land Use & Activity Review   8

Task 3 – Route & Schedule   14

Task 4 – Track-side Structures   16

Task 5 – Historic Resources   18

Task 6 – Trains   19

Task 7 – Track & Track Construction   21

Task 8 – Power Supply & Distribution   22

Task 9 – Origination and Termination Points   24

    a) Passenger Boarding   25
    b) Employee Distribution   25

Task 10 – Parking   26

Task 11 – Fare   26

Task 12 – Train Storage   27

Task 13 – Train Repair/Maintenance   28

Task 14 – Tunnel to Street Surface Interface   28

Task 15 – Traffic Considerations   29

EIS Analysis & Thresholds

Lot more details in description, what methods for analysis, and how to come to conclusions.

Section on specs and locations of pedestrian islands

Specify permits in permit section

Large scale maps with dimensions of streets

Street by Street {
- how wide easement would be
- what is street today
- what will street be with no project — Downtown Brooklyn Master Plan Urb. tran.
- what will be with project — Capacity of sidewalk and intersections
}

Land use: inventory all land uses within 400 feet of right of way (Noise, Air Quality, Exposures (Supply, distribution), hazardous waste/material, Fauna if any, Flora, public health)

1) Determine what existing conditions are
2) Future No Build — pick a date
   - what would world look like — metrotech, return Pierrpont
   - Economics Boro office will work with Economics division
   - provide logical projection on getting to no build

Get background, growth to form, Boro office

3) Future growth
   Comparison — showing effect on existing environment

4) Then compare future build effects on environment
   - This is potential impact

<u>I</u> consider each area and then propose study area
   propose 400 feet off right of way, give reason
   Ask UMTA for a copy of a certified EIS, such as Boston, Miami, Washington DC

<u>II Traffic</u> Propose C from letter Build year

If V/C 7.85 — has to incorporate Downtown Master Plan
need mitigation and you increase by more than 0.01 Signal Timing + Street widening
Double Parking
Conditions on Washington St
loss of parking spaces
bus stops should not compete with light rail stops

Assume all existing buses, trips, new people we there. Do not substitute for buses

Quantify beneficial
Impacts
Show capacities, V/c, numbers
to show improvement in traffic & air

Task		Page Number
Task 16	Noise	31
Task 17	Open Spaces Recreational Resources	32
Task 18	Air Quality	33
Task 19	Traffic Impact Analysis	35
Task 20	Transportation	46
Task 21	Construction	54
Task 22	Alternatives	57

Take measurements at desired locations and then a modeling process. Propose tables. Sample on a variety of scales, build for future. Calculate by hand for 3 sites. Propose dedicated lanes for LRV's. Calculate noise produced by LRT and noise levels & traffic at different speeds. Parsons has study on this. Impacts on receptors, parks, schools, hospitals.

Show all locations of kiosks and islands for loading — sidewalk capacity — Traffic flow
Analyze congested areas — bus stops at key intersections — places where double parking occurs — Islands only for passengers — no painted islands

Towing service to move stalled trolleys
What effect on schedule would a broken trolley have — Traffic impact what if track was blocked by truck or obstruction

Specify which Traffic lights would be triggered by trolley. Show effect of ripple effect. Will changes in signals occur at key intersects, will trolleys wait at others. Suggest where I want them tripped, study traffic & air quality impact + mitigation

Provide typical brake

Specify envelope for maximum size car

Current histories of operation what happens when car breaks down - Losses power in intersection

Hazardous materials

De Icing materials

Lubricants use collection disposal of

Any PCB's in motors

Alternative section

PROJECT DESCRIPTION

Background

The Brooklyn Heights/Cobble Hill section of Brooklyn is the home of the world's first subway tunnel, constructed in 1844 by the Long Island Rail Road. The structure was the first railroad tunnel built by the now standard "cut and cover method" of subway construction. As part of the first rail link between New York Harbor and Boston, the tunnel carried steam powered trains below Atlantic Avenue for 17 years until being sealed and abandoned in 1861. For over 100 years the tunnel lay dormant beneath Atlantic Avenue until its rediscovery in 1980 by a young Brooklyn resident, Robert Diamond. After extensive research, Mr. Diamond located the only existing entrance to the tunnel and began a series of exploratory ventures into this historic structure. With the assistance of local volunteers, Mr. Diamond constructed a 70-foot long accessway from a manhole through a backfilled section of tunnel near Court Street, making the tunnel's interior accessable from the street, revealing the one-half mile long, perfectly preserved structure.

From these beginnings, an exciting project which would improve transportation service into downtown Brooklyn has arisen.

After his rediscovery of the tunnel, Mr. Diamond formed a non-profit corporation to restore and reopen the tunnel. A detailed structural study was performed in 1985 by Singstad, Hurka Associates, and it was found that the tunnel was in perfect condition (see Appendix "A"), and did not require any repair work.

In 1986, Mr. Diamond's non-profit corporation, the Brooklyn Historic Railway Association, was awarded franchise rights to the tunnel by the Board of Estimate, renewable at ten-year intervals. In 1987, the City allocated $2.6 million in the adopted Capital Budget for the reconstruction of the tunnel's entrances which have been sealed since 1861.

The Project

Traffic congestion on arterial highways leading into New York City from Long Island is perpetually worsening, and if New York is to remain competitive with other urban areas which are centers of commerce, we must be able to provide adequate public transportation. It is vital for public transportation to make its resurgence now if New York is to maintain its position of preeminence. Many other major urban centers such as Seattle, San Diego, Buffalo, San Jose and Sacramento have already completed Light Rail systems. Still others have been begun in Los Angeles, Baltimore, Dallas, Houston and one is being planned for Miami. The city of Philadelphia is presently studying the feasibility of constructing an historic trolley route which would connect a large scale commercial and residential development at Penn's Landing with downtown Philadelphia.

Further, traffic in Manhattan and downtown Brooklyn on a business day is always at or near a gridlock condition. Travelling from Long Island to New York by car takes several hours during rush hours, even from relatively short distances. Also, New York City leads the nation in Carbon Monoxide pollution, and is third in Ozone pollution (see Appendix "B"). Figures are latest EPA data.

Long Island is well served by the Long Island Rail Road, however there is a large population segment on the Island which uses private automobiles to commute to the City because they refuse to ride the subway. The LIRR, because of historical and original design reasons, lacks a direct access route to the primary business areas of the City. The LIRR is not at capacity into New York City during the a.m. peak. In fact, the LIRR is considering reducing its fare to Atlantic Avenue in order to divert riders from Penn Station and attract new commuters working in Brooklyn (LIRR Memorandum dated June 10, 1988). Existing subway feeders to the LIRR in Brooklyn face overload conditions due to several major development projects.

It is the purpose of the Brooklyn City Railroad Project to remedy this situation by reactivating part of the LIRR's original route into New York City. We plan to attract this as yet untapped commuter segment and draw them to use mass transit. We also plan to divert local bus riders and automobile users, as well as midday shoppers, to the Light Rail service.

Development Program

The Brooklyn Historic Railway Association has applied for a franchise from the City of New York to construct and operate a Light Rail link from the LIRR terminal at Flatbush Avenue, Brooklyn, to Brooklyn Pier 6 (as well as other feeder routes), where ferries would take commuters to Pier 11 at Wall Street, Manhattan's East and West sides and Hoboken. It is even possible, from a technical point of view, to convey the Light Rail vehicles on the ferry, thereby eliminating all but one passenger transfer (from the LIRR to the Light Rail). The Light Rail trains would operate on a 5-minute headway during rush hours. The system capacity (using twenty PCC type Light Rail vehicles, and assuming a 5-minute system wide headway) is 4,248 peak hour passengers in one direction at the theoretical maximum load point. Total commuting time from Flatbush Avenue to Wall Street would be 18 minutes. The projected fare for the combined Light Rail/Ferry ride is $5.00 per round trip.

Public Actions

The Brooklyn City Railroad Project would require several discretionary public actions that are subject to the City's Uniform Land Use Review Procedures (ULURP) and which have triggered the need for this EIS: a franchise contract from the Board of Estimate to construct and operate this proposed system as per Section 374 of the NYC Charter, consent from the New York City Department of Transportation to lay and operate track within City streets, as well as consent to reopen the tunnel portals which would require the widening of two street segments along Atlantic Avenue, (review of this is already underway as part of Builder's Pavement Plan #BNP-88-262) and the change of legal grade elevations to accomplish the portal reopening. It would also be required that a provision for a Light Rail track be included in the Port Authority's plans for the widening of Furman Street as part of the redevelopment of Brooklyn Piers 1-6. Consent to operate the ferry link would also be needed from the NYC DOT's Office of Private Ferry Operations. Permission will also be needed from the NYC Parks Department to traverse Columbus Park, as well as the New York State Department of Parks for consent to traverse Empire State Fulton Ferry Park.

Legislative Requirements

A Draft Environmental Impact Statement will be prepared for the proposed project. This Statement will be prepared in conformance with the requirements of Executive Order No. 91, City Environmental Quality Review regulations dated August 24, 1977. It will contain the following:

1) A description of the proposed action, its purpose and need;

2) A description of the environmental setting;

3) A statement of the environmental impacts of the proposed action, including its short-term and long-term effects, and typical associated environmental effects;

4) An identification of any adverse environmental effects which cannot be avoided if the proposed action is implemented;

5) A description and evaluation of reasonable alternatives to the action that would achieve the same or similar objectives;

6) An identification of any irreversible and irretrievable commitments of resources that would be involved in the proposed action;

7) A description of mitigation measures proposed to minimize adverse environmental impacts;

8) A description of any growth inducing aspects of the proposed action;

9) A discussion of the effects of the proposed action on the use and conservation of energy; and

10) A list of underlying studies, reports and other information obtained and considered in preparing the statement.

TASK 1

Project Scope

The scope of this project is two-fold; first, as an internal transit mode for downtown Brooklyn, and second as a regional transportation link. Downtown Brooklyn is experiencing an economic rebirth. Several major institutional and commercial developments, totalling over 20.5 million square feet, are planned. There are also several Brooklyn cultural institutions within the Light Rail corridor, such as the Brooklyn Museum, Botanic Gardens, Brooklyn Academy of Music, etc. All of these

sites would be tied together with eachother, as well as
Manhattan and New Jersey, by the Light Rail and Ferry system in
a way which would not increase traffic congestion and air
pollution. In addition, the trolleys would themselves become
tourist attractions, as are the cable cars in San Francisco.
Since the Light Rail route would operate via the Atlantic
Avenue tunnel, there is also the added dimension of a true
historic attraction along the route.

An additional regional aspect of the project is the development
of large numbers of homes in Suffolk County. New Jersey has
mounted a full scale campaign to attract suburban dwellers to
similar developments along the west shore of the Hudson River,
as well as Monmouth and Mercer counties. They are boasting
easy access to New York City by public transportation. In
fact, the New Jersey DOT has already begun engineering plans
for an extensive Light Rail system which would link many of
their suburban developments to a Ferry service to Manhattan.
It would be to New York City's advantage to prevent the
emergence of a "transportation gap".

TASK 2

Land Use and Activity Review

Downtown Brooklyn includes the traditional Central Business District and Civic Center of the Borough of Brooklyn (Kings County), as well as the adjoining Brooklyn Heights neighborhood and waterfront. This represents one of the largest downtown Central Business Districts in the United States, an important center of Federal, State and City offices and a center of higher education. Downtown Brooklyn and the residential communities around it have been in decline for over a century since the termination of Brooklyn railroad service in 1861, and has been traditionally overshadowed by Manhattan, both economically and culturally. However, since the 1960's Greater Downtown Brooklyn has shown signs of revival both because of the renovation of brownstones by a rapidly growing upper middle class community, and due to the planning and construction of a number of Downtown redevelopment projects like Fulton Mall, the Albee Square Mall, Morgan Stanley, renovation of landmarks, urban renewal projects at Cadman Plaza, Fort Greene, the Long Island Rail Road Atlantic Terminal, Fulton Landing, the Brooklyn Academy of Music, the Botanic Gardens, etc. Because of its proximity to Long Island suburbs, low real estate values and under-used prime location land, Downtown Brooklyn is now perceived as ripe for expansion and redevelopment. In 1980, it was estimated that there were approximately 8.1 million square feet of office space in Downtown Brooklyn, as well as over 2 million square feet of rental space. An extra 5 million square feet of space is now almost totally occupied in Fulton Landing. Fulton Landing also houses the New York State Department of

Labor.

Government buildings account for almost half of all office space. Public institutions in downtown Brooklyn include the Brooklyn Central Post Office, the Federal Building, the City Board of Education, the Transit Authority, the Fire Department and the Worker's Compensation Board. Presently, the Brooklyn CBD supports over 80,000 jobs, half of which are in the public sector.

Downtown Brooklyn is also a center for public utilities - the Brooklyn Union Gas Company, New York Telephone, Consolidated Edison, as well as financial institutions (fourteen commercial and six savings banks). It contains educational institutions, including Long Island University, Polytechnic University of New York, Brooklyn Law School, New York City Technical College, Pratt Institute, Saint Francis College, the Institute of Design, Brooklyn Technical High School, Westinghouse Vocational and Technical High School, Packer Collegiate Institute and Saint Joseph's College/High School.

Figure 1 shows the existing land use in downtown Brooklyn. Of note on this map are the following: 1) the central commercial core of downtown Brooklyn, centering on Fulton Mall; 2) the Civic Center complex just to the west of this between Court and Jay Streets; 3) the concentration of colleges and public agencies in and around the central downtown area; 4) the lack

DOWNTOWN BROOKLYN

FIGURE 4   MAJOR OFFICE SPACE FACILITIES

DOWNTOWN BROOKLYN

FIGURE 5 DEVELOPMENT SITES

NUMBER	NAME	APPROXIMATE SQ.FT.
1.	ATLANTIC TERMINAL BACK OFFICE DEVELOPMENT	4.6 MILLION
2.	YWCA REHABILITATION-STATE OFFICES	300,000
3.	BARTON FACTORY CONVERSION	300,000
4.	GILDNER BUILDING CONVERSION	100,000
5.	OFFICE REHABILITATION (2 SITES)	62,000
6.	MCCRORY BUILDING RENOVATION	80,000
7.	OFFICE DEVELOPMENT OVER A & S STORE	125,000
8.	RENOVATION OF EXISTING OFFICES	40,000
9.	PROPOSED MIDDLE/UPPER INCOME HOUSING	—
10.	RENOVATION-OFFICES	40,000
11.	RENOVATION-OFFICES	40,000
12.	OFFICE CONSTRUCTION-5 FLOORS	360,000
13.	OFFICE CONSTRUCTION-10 FLOORS	600,000
14.	HOTEL DEVELOPMENT OVER GARAGE	—
15.	81 WILLOUGHBY RENOVATION	80,000
16.	METROTECH	2.9 MILLION
17.	OFFICE TOWER-10 STORIES	460,000
18.	OFFICE/RETAIL BUILDING	80,000
19.	101 CLINTON REHABILITATION	65,000
20.	360 FURMAN STREET (MIXED USE)	816,000
21.	PORT AUTHORITY WATERFRONT	(51 ACRES)
22.	WATCHTOWER SOCIETY RESIDENCES	(30 STORIES)
23.	FULTON LANDING	2.2 MILLION (MIXED USE)

URBITRAN ASSOCIATES, INC

DOWNTOWN BROOKLYN

FIGURE 6 APPROXIMATE SQUARE FOOTAGE DEVELOPMENT AREAS

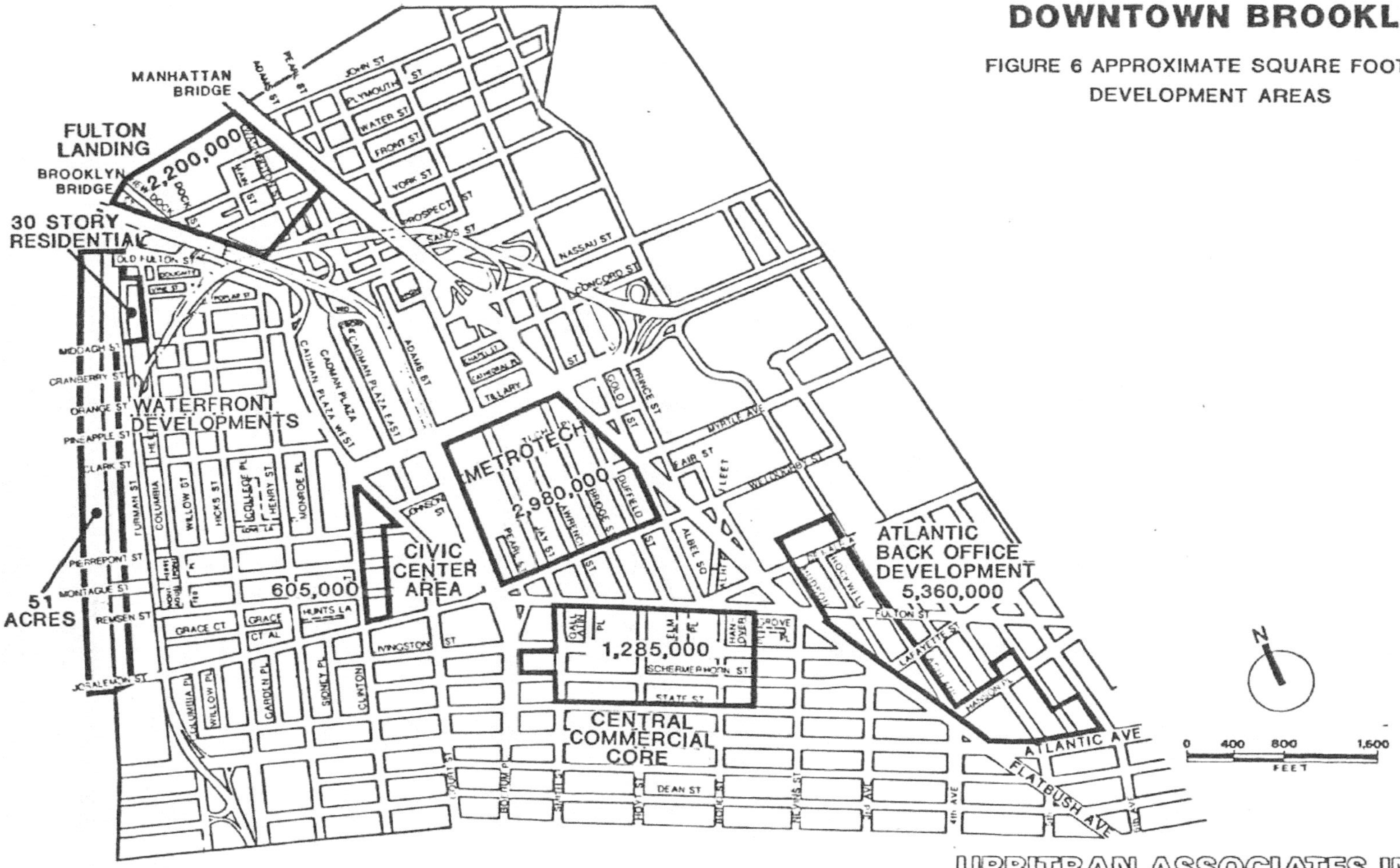

URBITRAN ASSOCIATES, INC

DOWNTOWN BROOKLYN

FIGURE 7
APPROXIMATE SQUARE FOOTAGE – COMBINED EXISTING/PROPOSED COMMERCIAL DEVELOPMENT

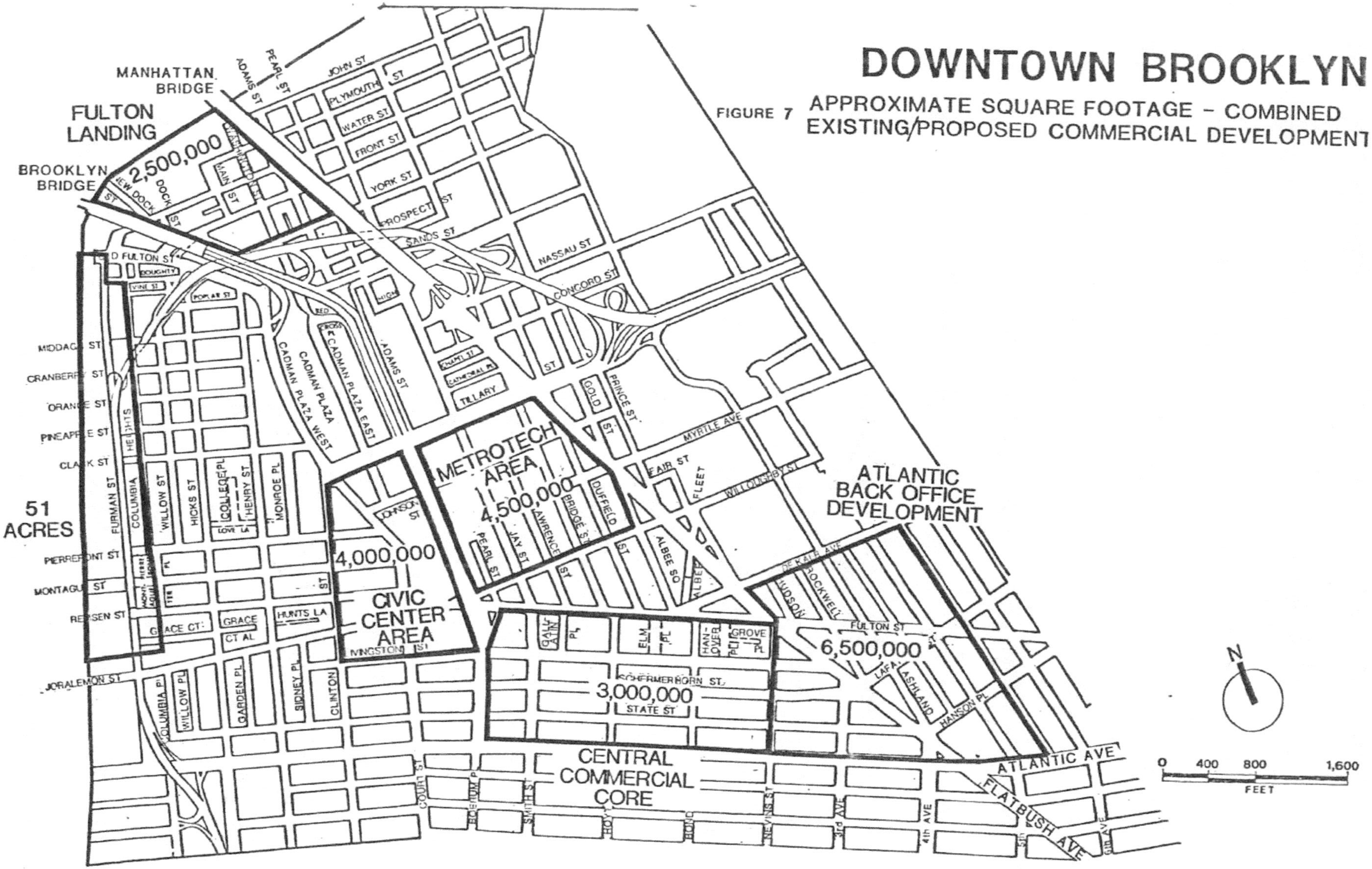

- Fulton Landing: 2,500,000
- Metrotech Area: 4,500,000
- Civic Center Area: 4,000,000
- Atlantic Back Office Development: 6,500,000
- Central Commercial Core: 3,000,000
- 51 Acres

URBITRAN ASSOCIATES INC

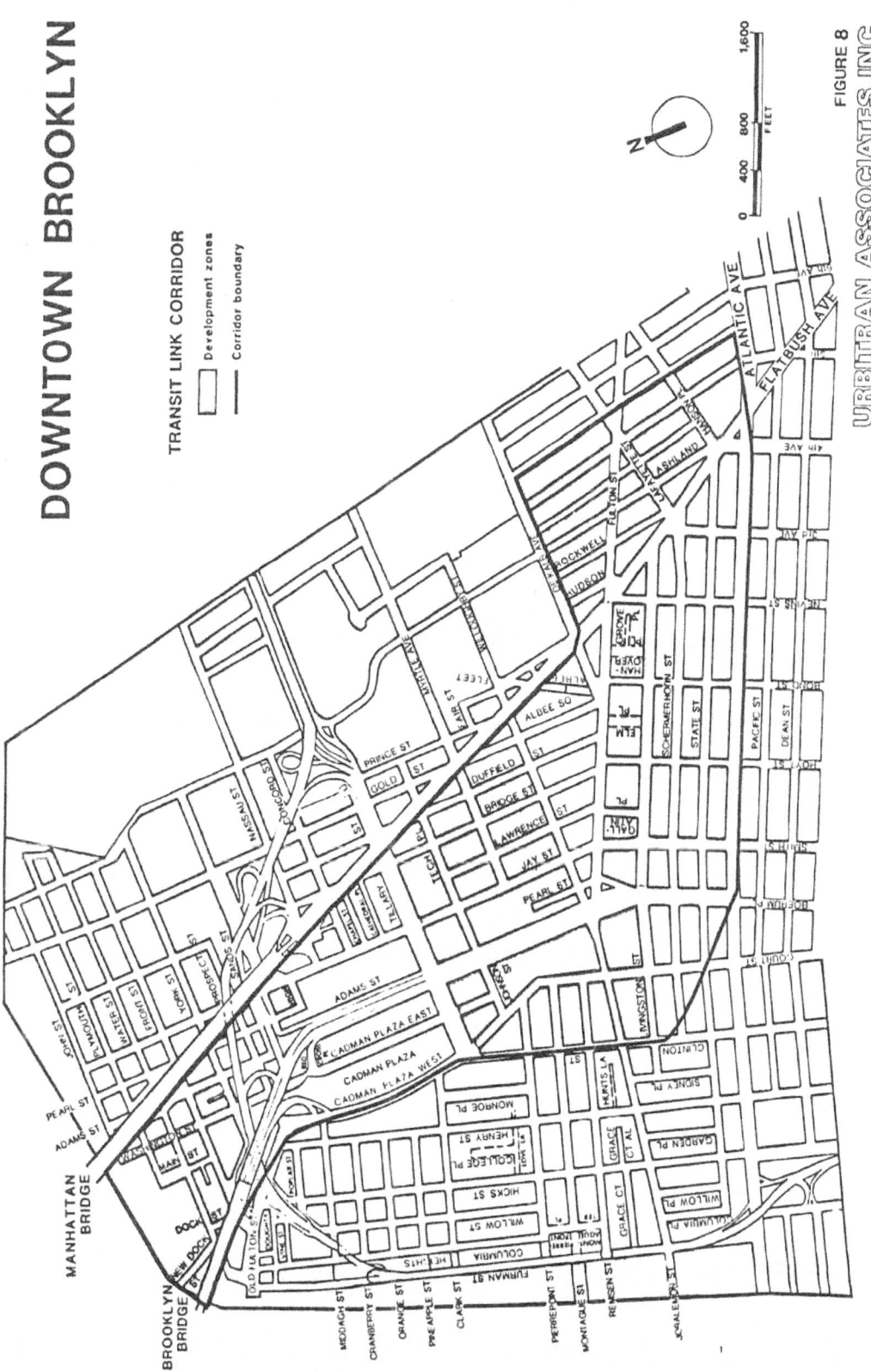

DOWNTOWN BROOKLYN

TRANSIT LINK CORRIDOR
- Development zones
- Corridor boundary

FIGURE 8

URBITRAN ASSOCIATES, INC

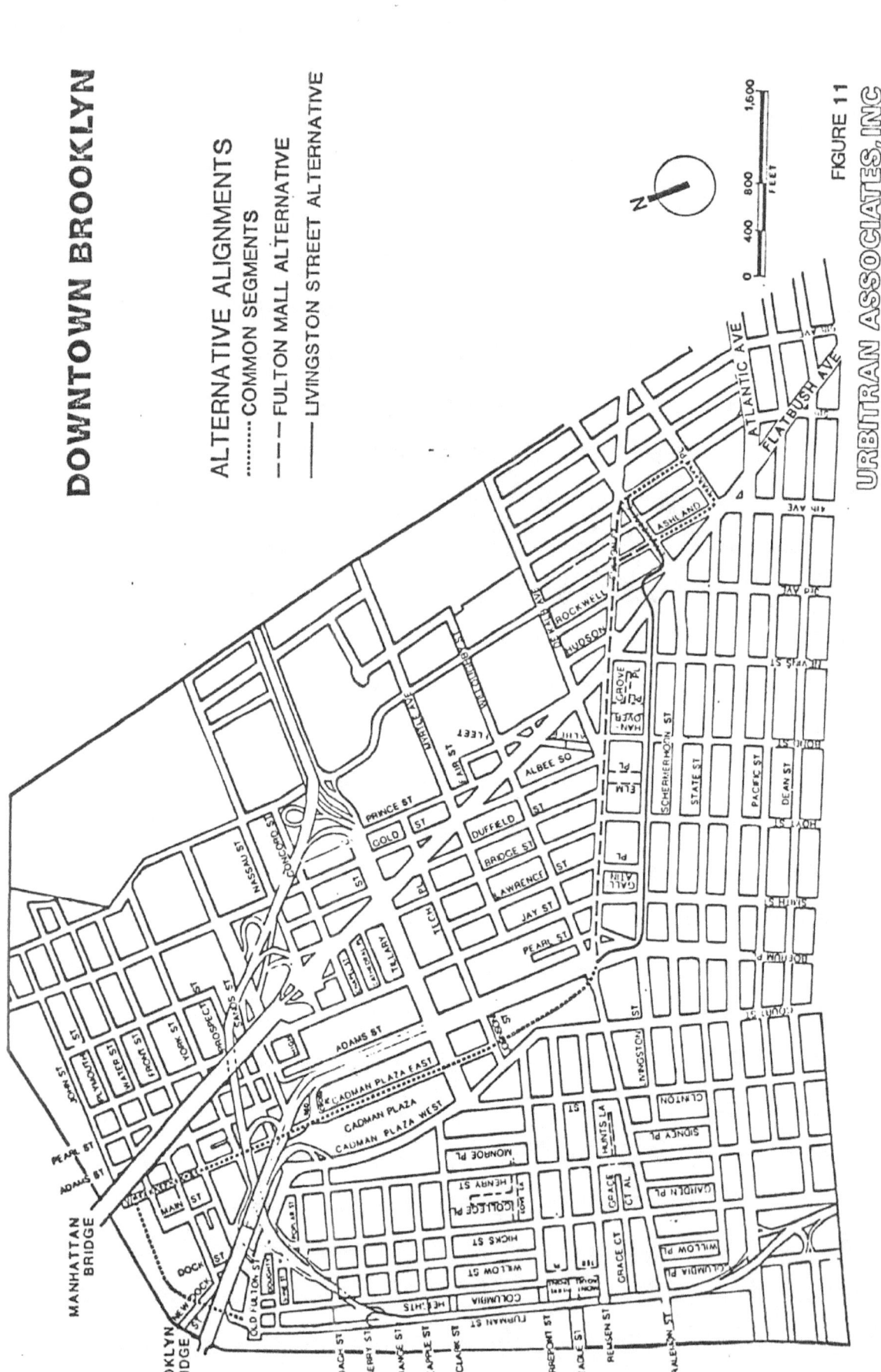

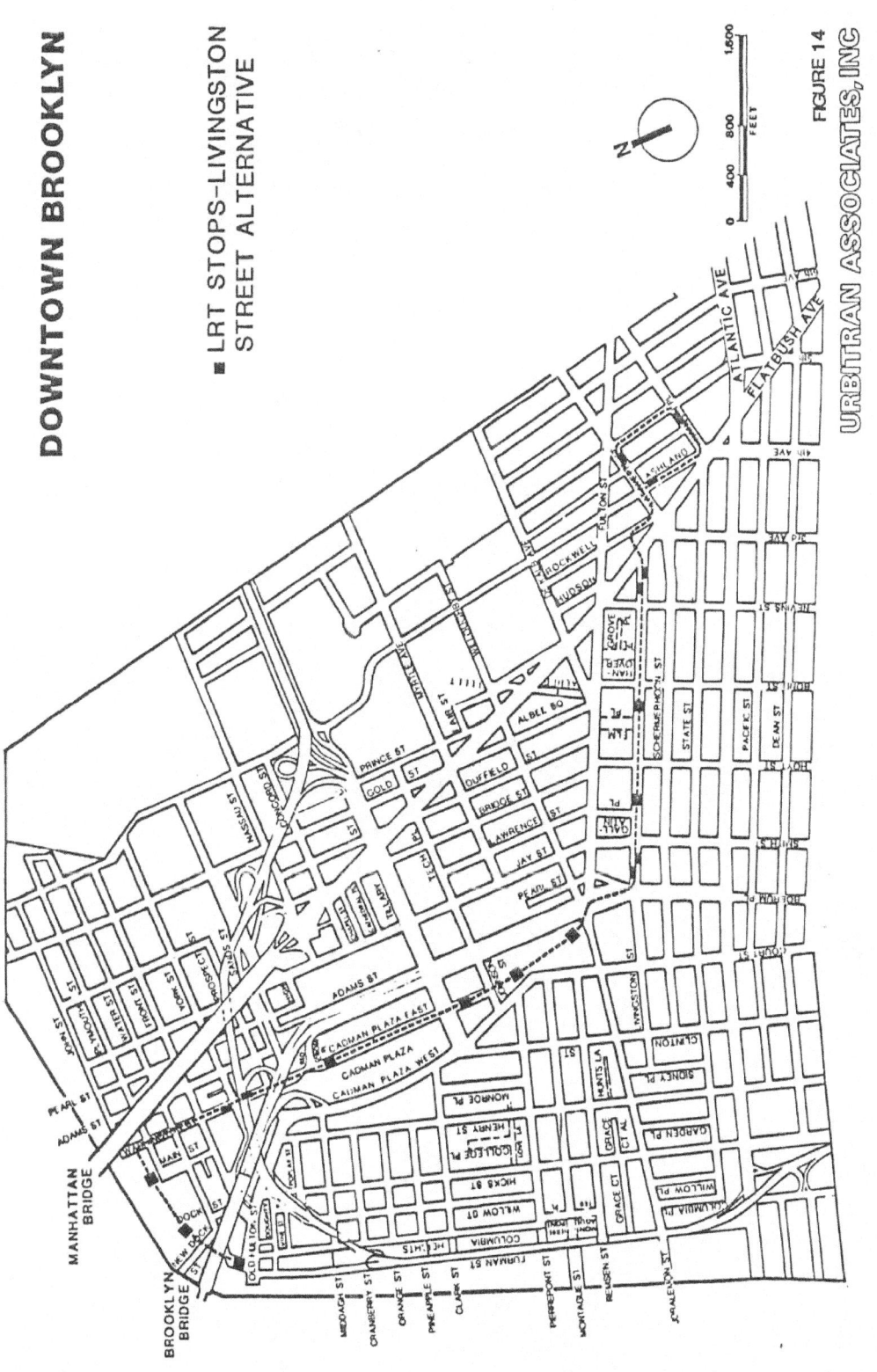

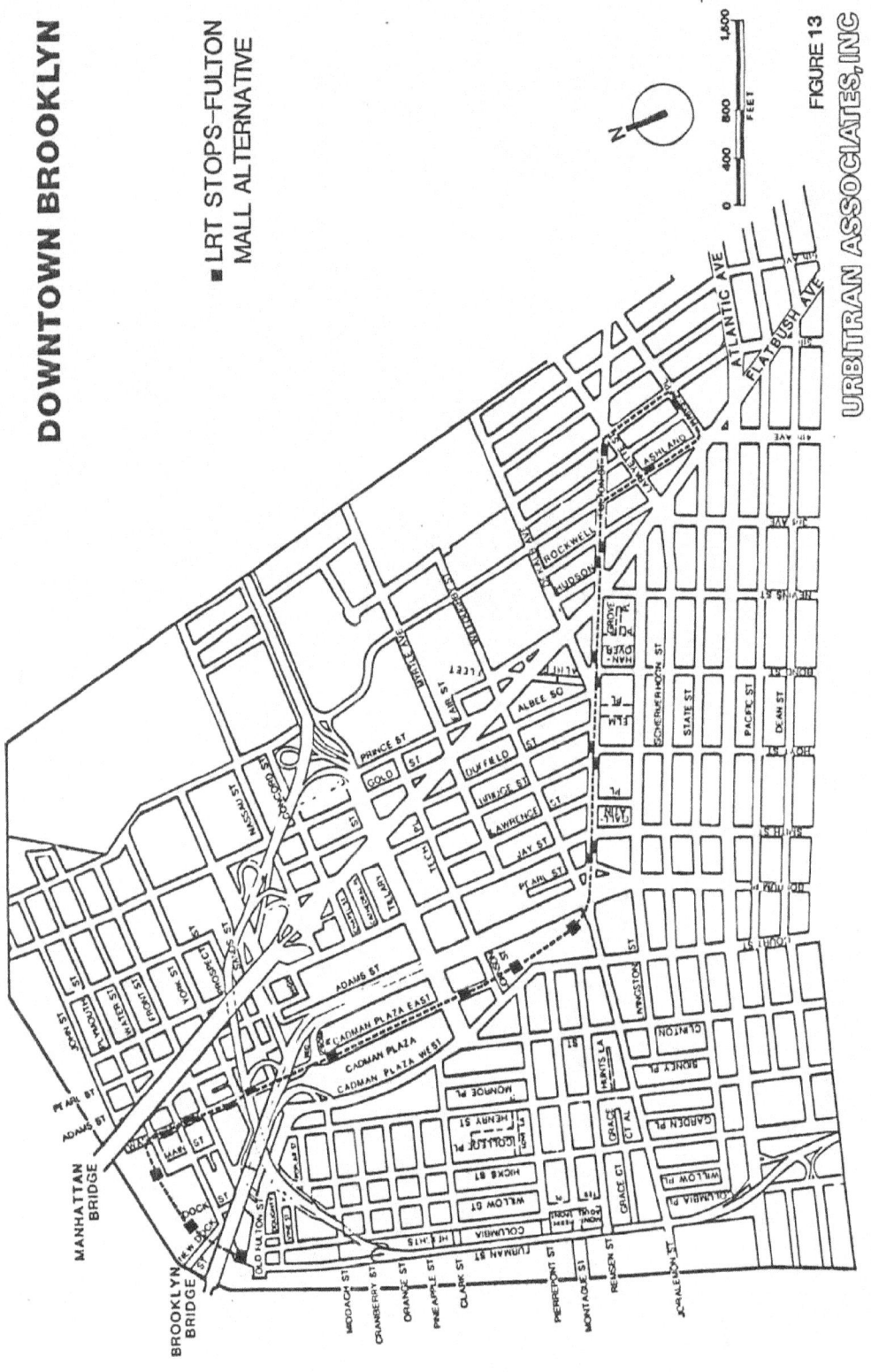

DOWNTOWN BROOKLYN

■ LRT STOPS–FULTON MALL ALTERNATIVE

FIGURE 13

URBITRAN ASSOCIATES, INC

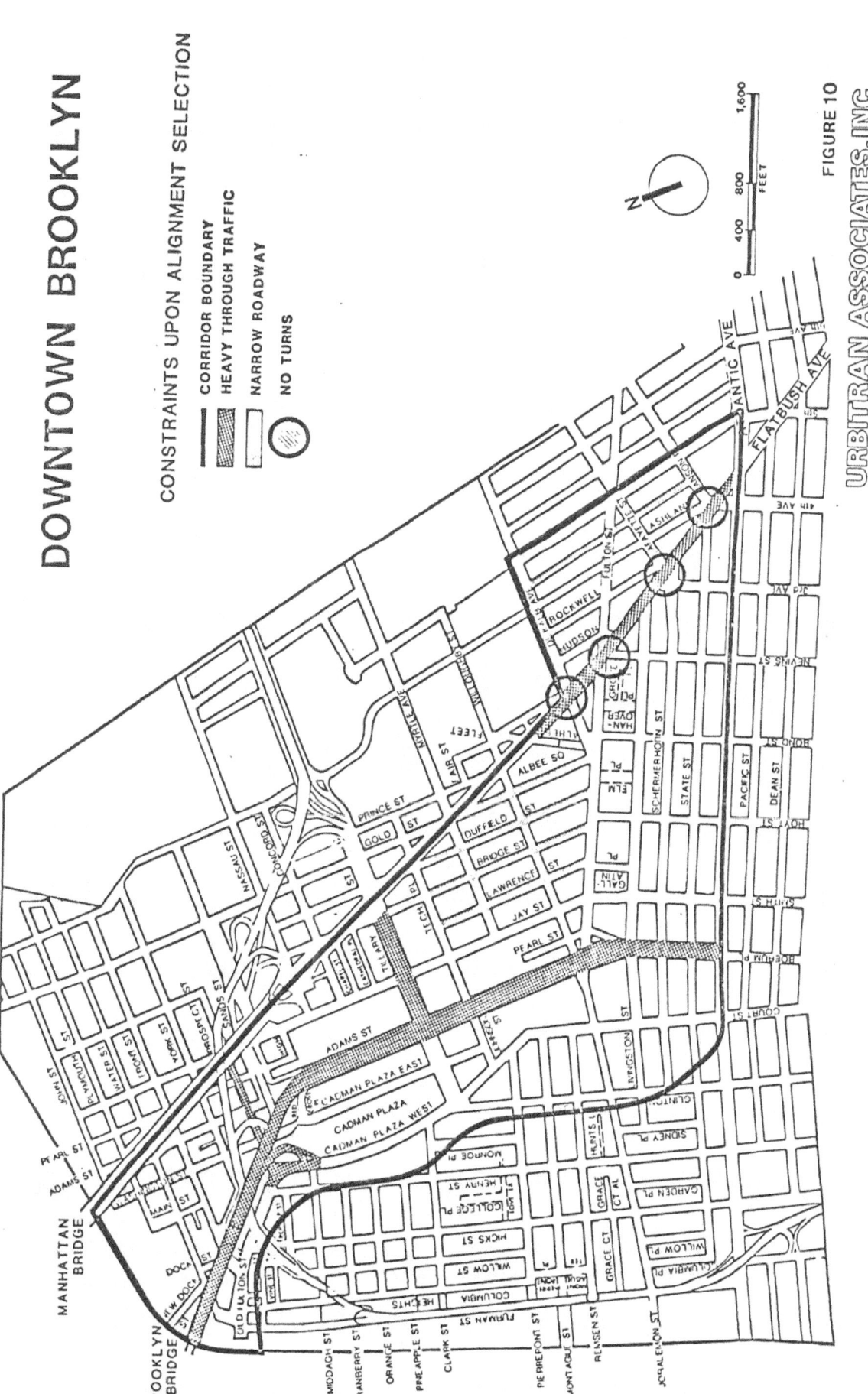

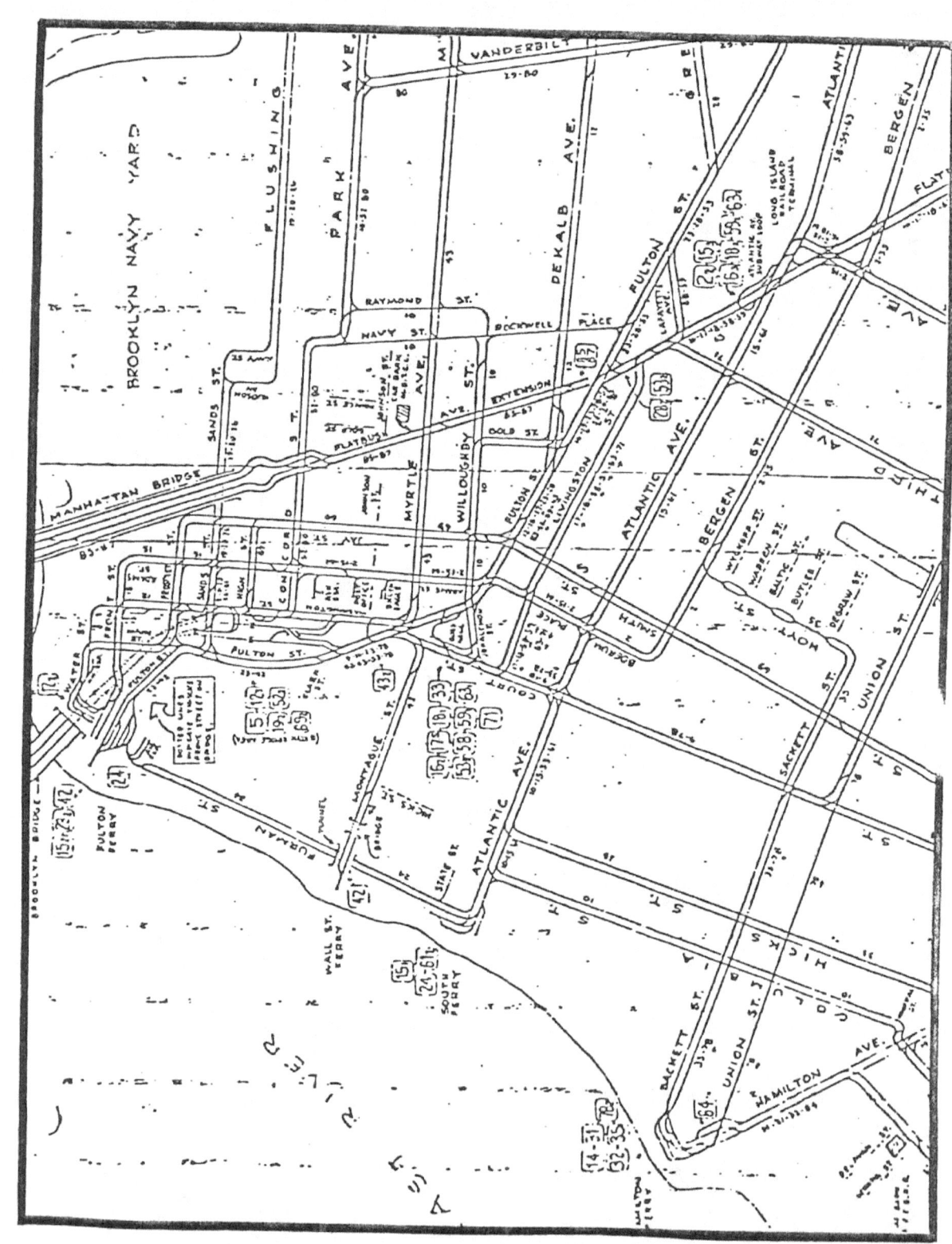

FIGURE 9
1914 STREETCAR PLAN–DOWNTOWN BROOKLYN

URBITRAN

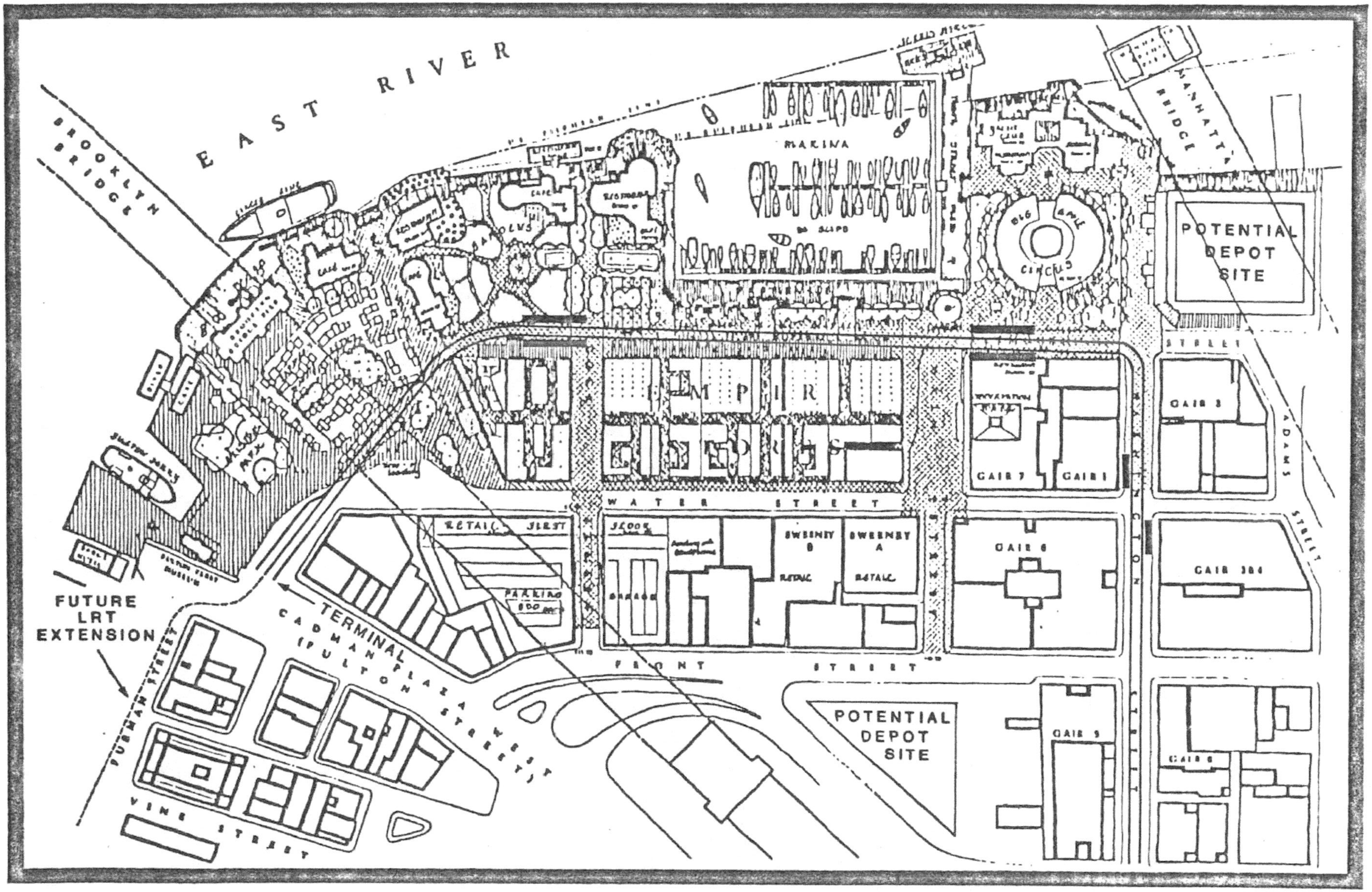

FIGURE 12
LRT ALIGNMENT WITHIN THE FULTON LANDING, DEVELOPMENT

URBITRAN

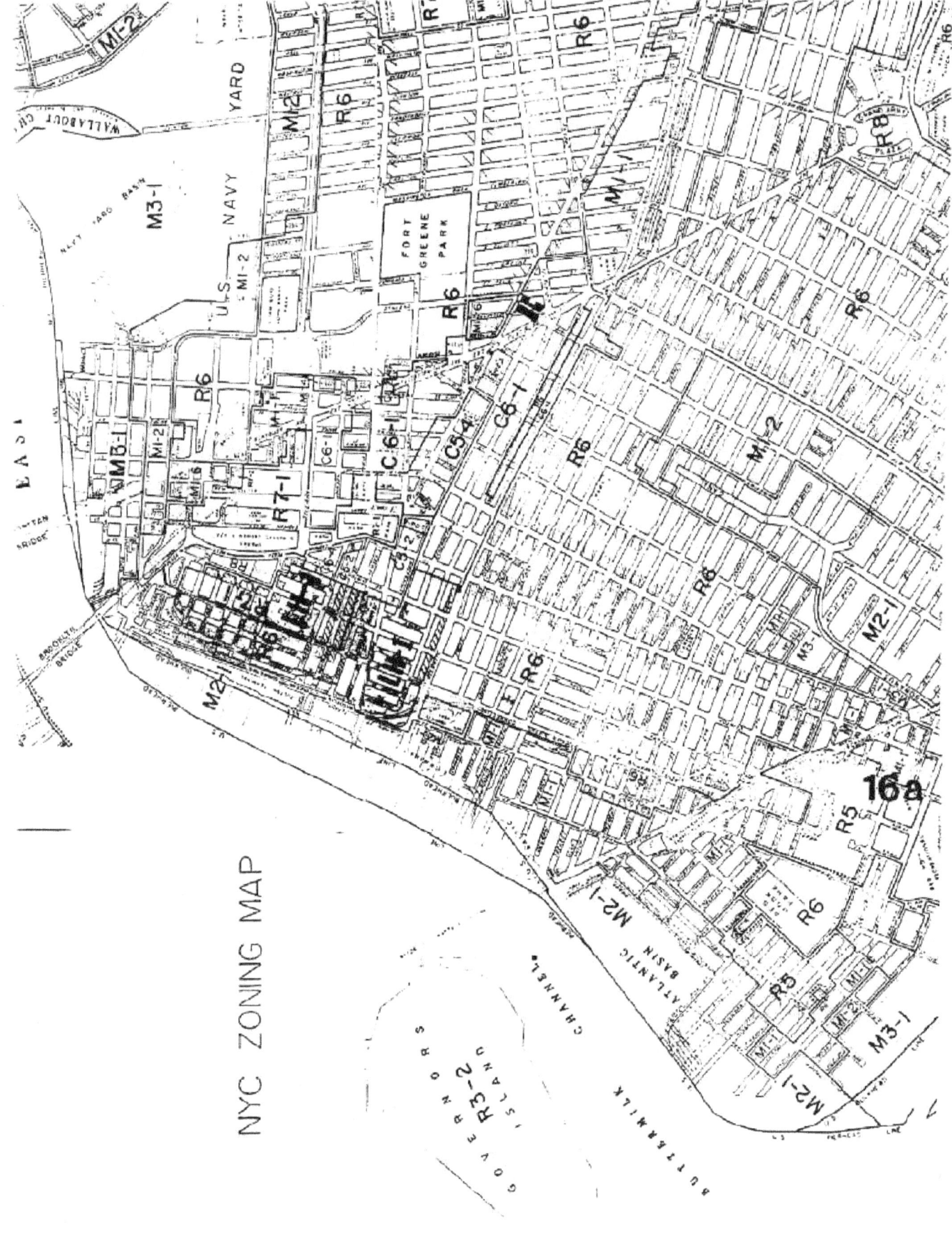

NYC ZONING MAP

of intensive land uses throughout much of this area around the LIRR terminal area, and in the area around the Brooklyn and Manhattan bridges; and 5) the presence of the unused waterfront piers to the west of Brooklyn Heights.

The Polytechnic University of New York's (PINY) Metropolitan Technology Center (MetroTech) plan is a joint PINY/City development project to create a Research and Development and educational center totalling 4.23 million square feet. Atlantic Terminal/Brooklyn Center will contain a total of 3.38 million square feet of commercial space, as well as 700 residential units.

No Build Conditon, 1993:

While downtown Brooklyn will experience an unprecedented renaissance due to the major redevelopment projects which will be on line at this time, the area will be facing an overload breakdown of its conventional transportation systems due to traffic congestion and overloaded subways. While generating new jobs and economic expansion, projects like MetroTech and Atlantic Terminal/Brooklyn Center will overload all conventional forms of transportation in downtown Brooklyn, with the attendant increases in air pollution, noise and traffic congestion. Access to downtown Brooklyn will become difficult and unpleasant.

Build Condition, 1993:

The proposed Light Rail project will provide easy and pleasant access to and within downtown Brooklyn and Red Hook. By rechanneling people from existing bus routes as well as automobiles, traffic and air pollution in Brooklyn will be reduced. The area will be a more congenial place in which to live and work.

Socio Economics:

The Light Rail project will provide a new mode of transportation, which is clean and inexpensive to serve the expanded workforce and new residents resulting from development projects already planned, approved and soon to begin construction in downtown Brooklyn.

Secondary Impact on Displacement:

The incremental enhancement of development along the Light Rail route caused by the Light Rail will be minor when compared to development encouraged by such approved major projects as MetroTech and Atlantic Terminal Brooklyn Center. For example, the entire Light Rail System is encompassed by the Primary Trade Area of the Atlantic Terminal/Brooklyn Center project alone. See Figure IIE-2.

With regard to Red Hook, existing economic pressures already at work (which can be demonstrated by real estate transactions) have initiated a pattern of change in this neighborhood. The Light Rail might speed up change, but is not the cause of the change. The improved transportation will assist low income residents of the neighborhood.

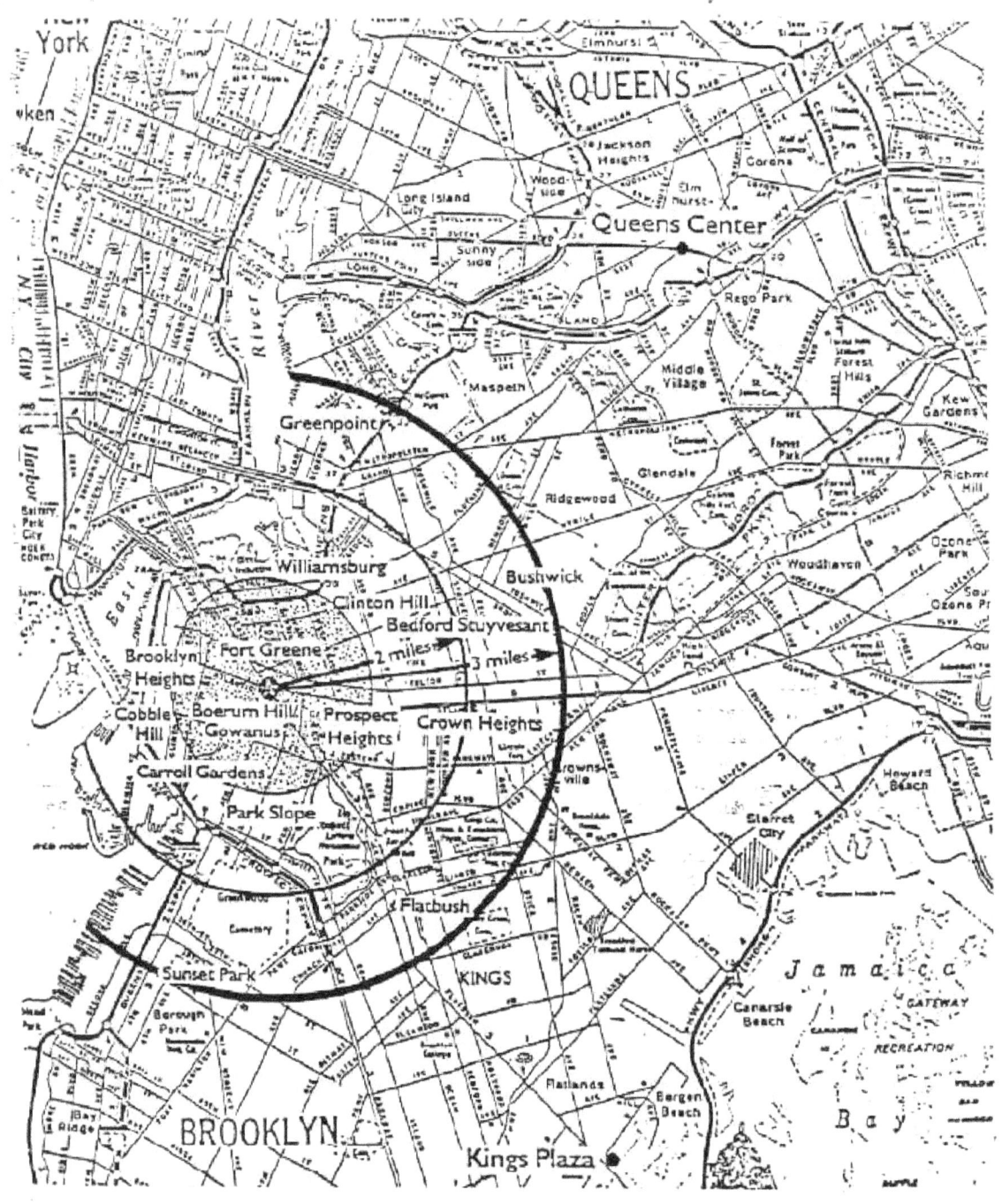

Shoppers' Goods Analysis Trade Area
Figure IIE-2

- Primary Trade Area
- Secondary Trade Area
- Project Site
- Neighborhood Retail Analysis Trade Area

Atlantic Terminal/Brooklyn Center

TASK 3

Route and Schedule

The routes as defined are designed to facilitate the two basic functions of the proposed Light Rail system. First, as a connecting web for downtown Brooklyn residents, development projects, businesses, cultural attractions and institutions ; and second, to serve as a regional link between the LIRR's Atlantic branch, Manhattan and New Jersey. The route of the system is indicated on Attachment "A".

The basic east-west surface corridor proposed is Fulton Mall because it is closed to automobile vehicular traffic, and its proximity to the central commercial core area. The main route then turns south along Boerum Place, and then west into the Atlantic Avenue Tunnel, emerging to grade at the waterfront at Pier 6, then turning north along Furman Street to Fulton Landing, then south again to reconnect with the main line at Boerum Place. There are also two feeder branches, one into Red Hook and the other to Eastern Parkway. The highest level of service will of course occur on the main line from Atlantic Terminal to Pier 6's ferry connection.

Number of Trains and Schedules

The total number of cars to be used is anticipated at 20. During peak hours some cars will be used in multiple-unit (M.U.) operation. The initially proposed service frequencies for the main route (pending a marketing and ridership analysis) is as follows:

Time Period	Weekdays	Weekends
6:00 a.m. – 10:00 a.m.	12 Trains/hr.	8 Trains/hr.
10:00 a.m. – 4:00 p.m.	12 Cars/hr.	6 Cars/hr.
4:00 p.m. – 7:00 p.m.	12 Trains/hr.	8 Trains/hr.
7:00 p.m. – 12:00 a.m.	6 Cars/hr.* 10 Cars/hr.**	6 Cars/hr.* 10 Cars/hr.**

* = Winter, ** = Summer

The service level of the branches will have to be determined subject to a detailed Market Survey. Service on the feeder branches will be by single car units. At this time service between the hours of midnight and 6:00 a.m. is not anticipated, it is stressed that the exact schedule will be determined by a Market Study. Trains on the main route would run express during peak hours. Additional trains may be phased in over time, as required. Local trains will pick up passengers on demand. Each stop will be less than 8 seconds. Local trains are defined as non-peak hour runs.

TASK 4

Track Side Structures

The largest track side structure considered is a re-creation of the original Atlantic Avenue Ferry Terminal which would be located near its original site, near Pier 6. See Attachment "B" for a schematic plan of the site. Also, see Attachment "C" for a drawing of this structure. All other "stations" would

1" = 200'

FURMAN ST.

NEW FURMAN ST.

atlantic ave.

ferry house/ ticket sales

PIER 6

FERRY

PIER 7

ATTACHMENT 'B'

Note: the City of Brooklyn may have been granted a "water grant" by the State legislature in the late 1830's. The lot requested was the width of Atlantic Ave + 50' on each side from water line to pierhead line

ATTACHMENT C

City of New York may own this lot now by success[ion]

Source: Stiles History of Bklyn V 3 1869 p 557

consist of loading islands. These would generally be 6 inches high, 5 feet wide and 150 feet long. For protection from the elements and approaching traffic, they would be equipped with simple but attractive shelters. See Diagram 1 for one possible design. The spacing between "stations" would vary from 500 to 1300 feet on the "local" stops. Some trains would skip stops during peak hours.

DIAGRAM 1

SHEET 1 OF 3

Notes:
1. For Section A-A, see Sheet 2.
2. For tunnel ventilating structure, see Sheet 3.

Legend:
——— Portion of tunnel presumed demolished
– – – Proposed entrance structure
▓▓▓ Existing tunnel structure

• BHRA •
Atlantic Ave. Tunnel
Entrance Concept
•
Preliminary Design
Plan & Profile

S.J.S. 11/83

PLAN
1" = 40'

- Proposed tunnel entrance
- Municipal Parking Garage building line
- Sidewalk
- East portal 115'±
- COURT ST.
- Earth fill
- Access MH
- ATLANTIC AVE.

PROFILE
Scale: H: 1"=40' V: ⅛"=2'

- Street surface
- East stone block bulkhead
- Existing earth fill
- 4' wide BHRA trench
- Concrete bulkhead
- Fill from BHRA trench
- Wood stair
- Access opening
- 1500'± clear to west stone block bulkhead

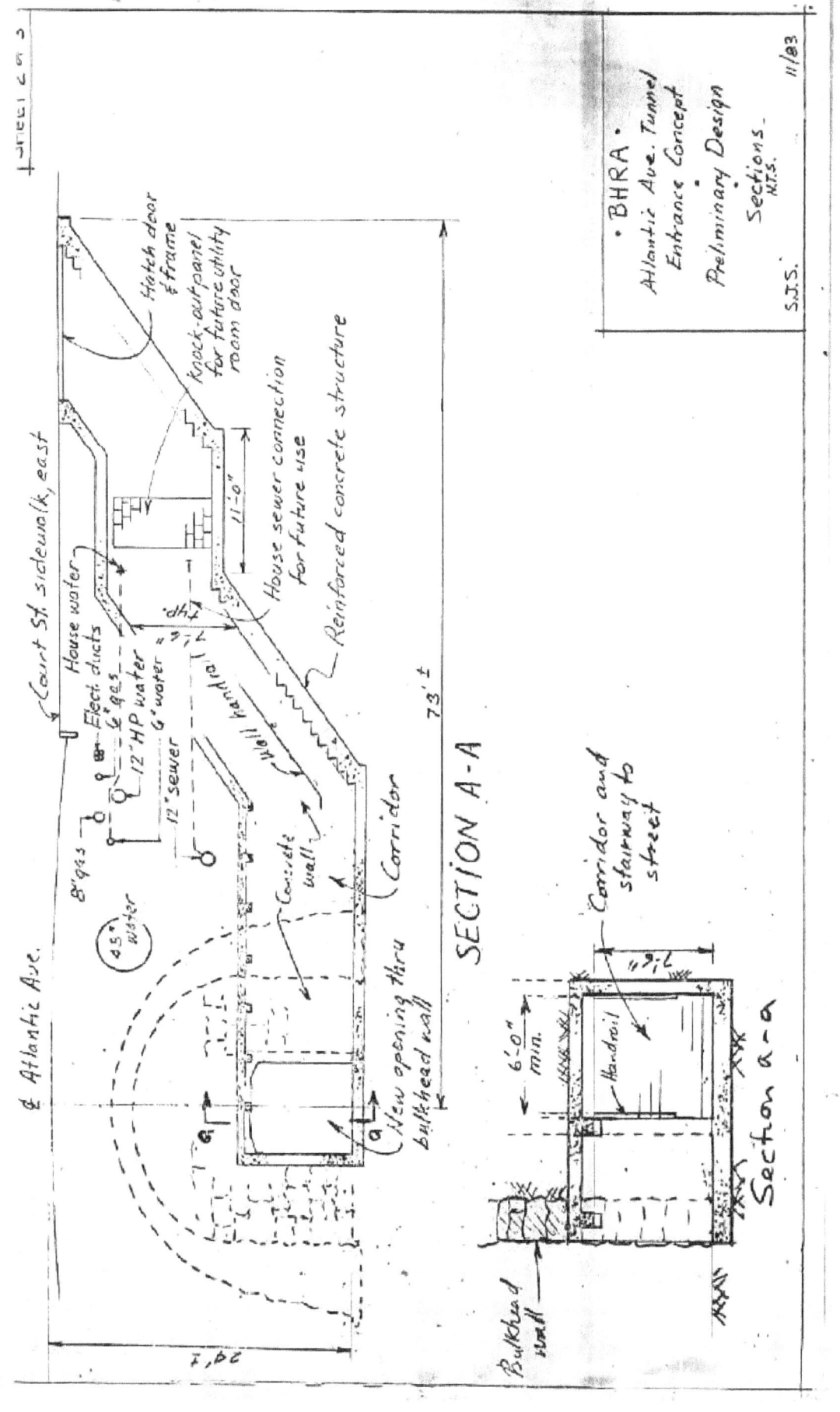

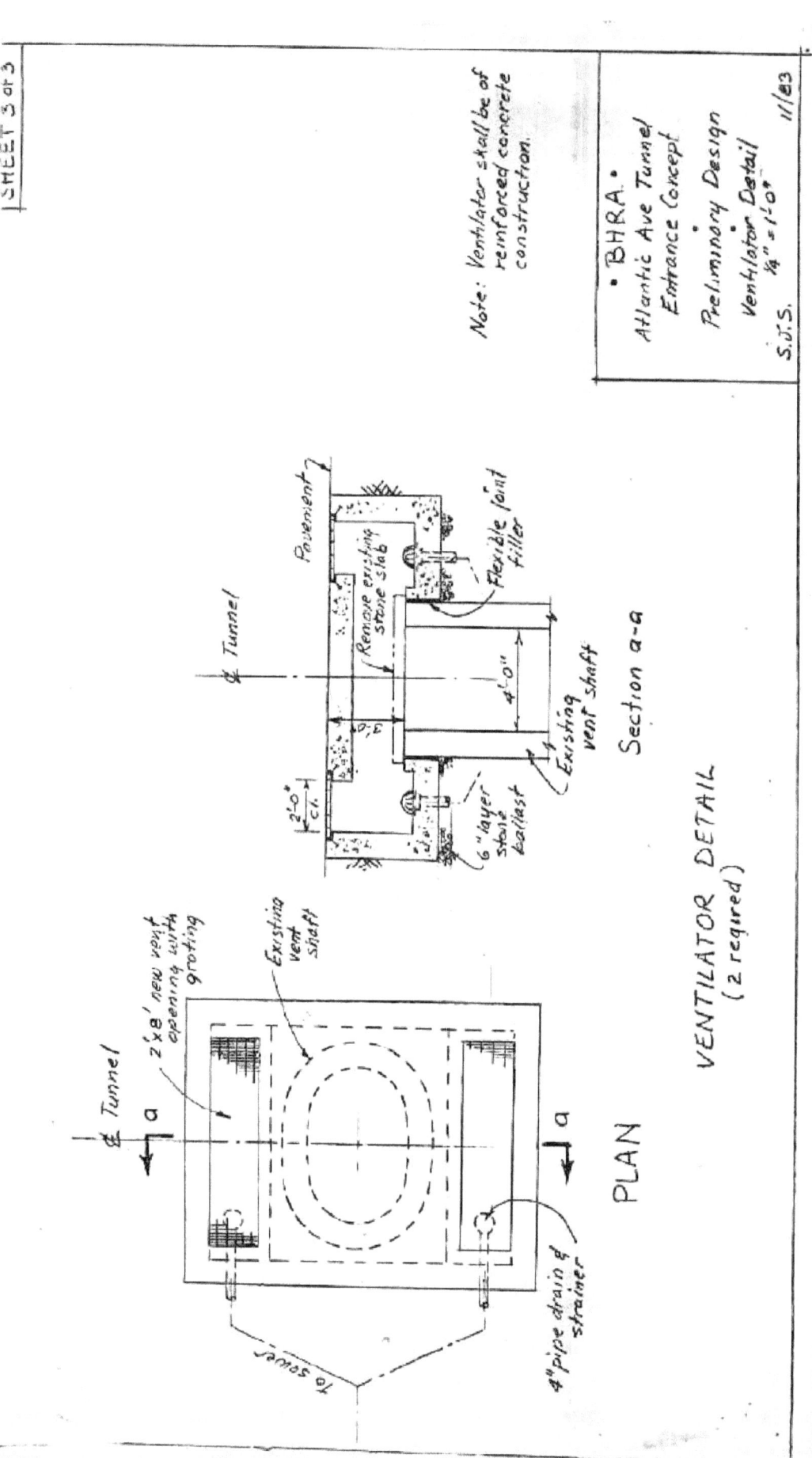

TASK 5

Historic Resources in Proximity of Light Rail System

1) The Atlantic Avenue Tunnel

2) Brooklyn Heights Bluff, site of the first military action of the newly formed U.S. Army in the Revolutionary War (August 27, 1776, Battle of Long Island, Washington's retreat to New York. One of the most decisive battles of the war).

3) Fulton Ferry, site of the first steam powered ferry boat operation.

4) Fulton Landing, site of the original Breukelen Village.

5) Brooklyn Heights, Cobble Hill and Fort Greene historic districts.

The Light Rail system does not pass directly through any

Metrotech Environmental Impact Statement

Historic Structures and Districts in Project Vicinity

EXHIBIT III-34

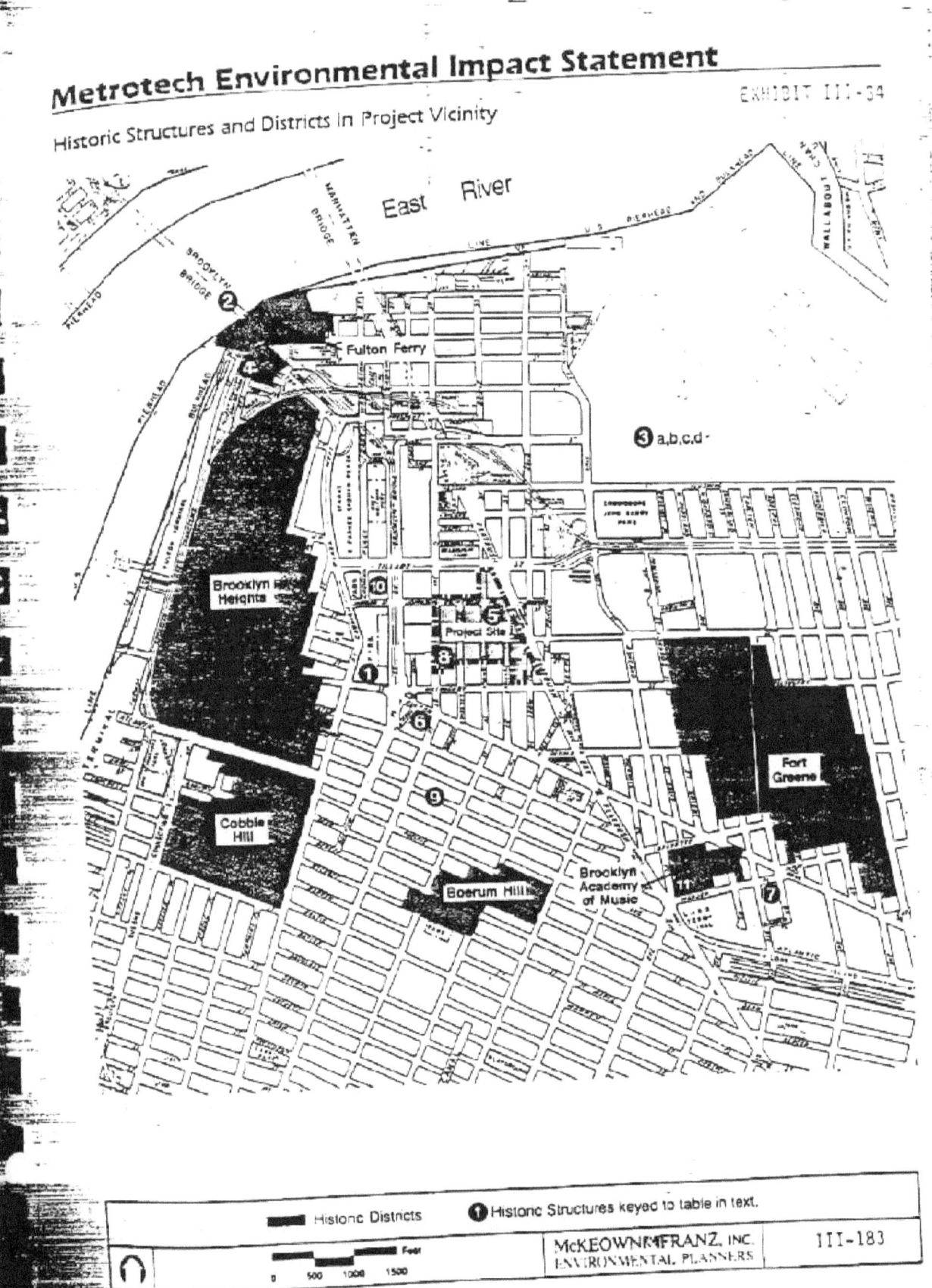

Historic Districts • Historic Structures keyed to table in text.

McKEOWN·FRANZ, INC.
ENVIRONMENTAL PLANNERS

III-183

13.41 HISTORIC DISTRICTS

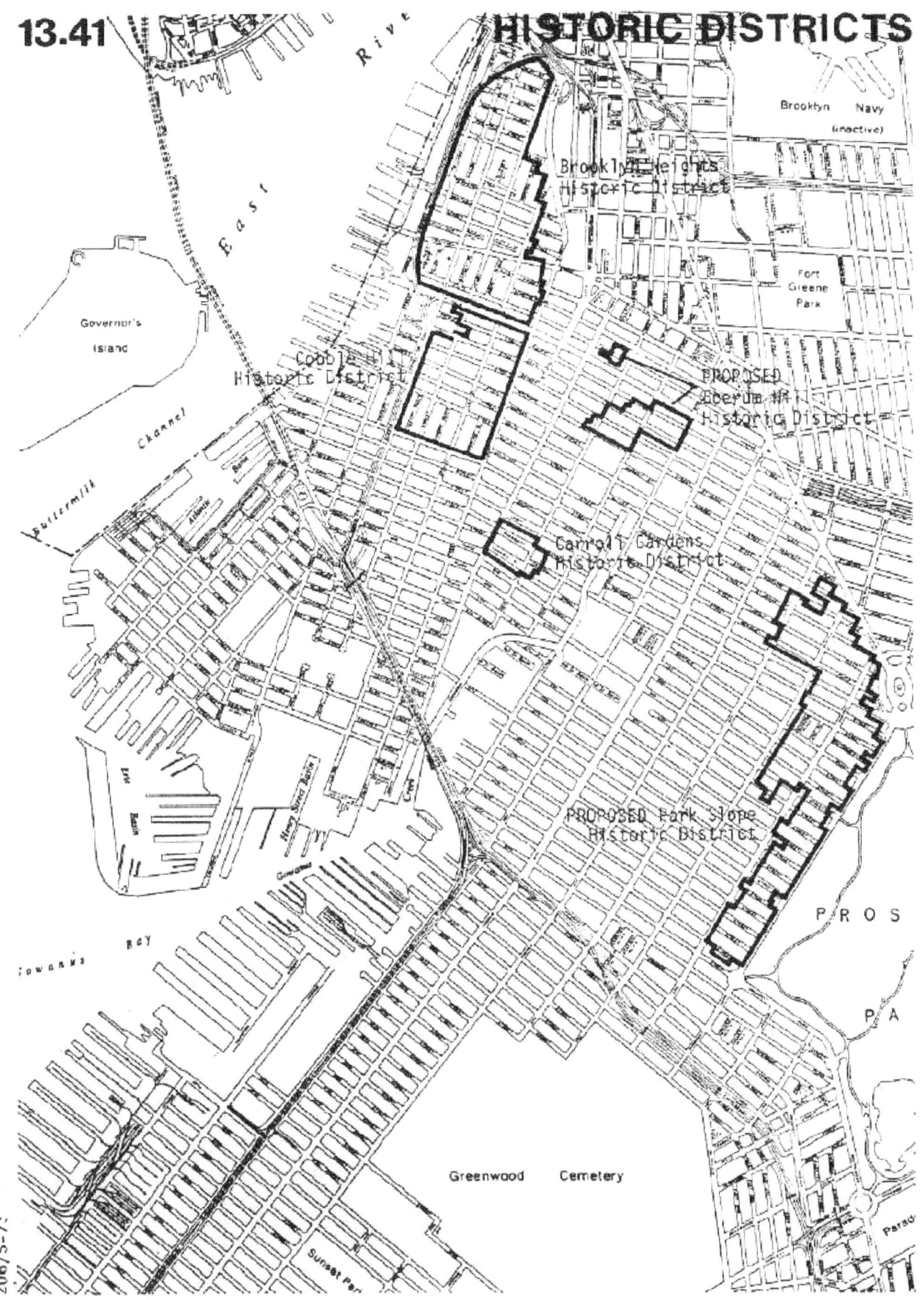

Source: Community Board 6 Planning Book

historic district. The maximum depth of excavation for track
installation is three feet.

TASK 6

Trains

An LRV can carry up to 150-180 persons at crush loads. The
average length of new LRV is 70 feet, older LRV's start at
50-foot lengths. The average width is about 8 1/2 feet.
Average height is about 12 feet. See Attachment "D" for
drawings of LRV's. In downtown use in Toronto and
Philadelphia, the average street speed for LRV's is 9-10 MPH.
Street running with adequate traffic controls is found commonly
in Europe. A Study by Taber and Lutin of the Toronto
streetcars found that 90 percent of total time delay for the
TTC cars were from boarding/alighting and traffic signals.
Actual congestion delays due to traffic were small. To
accomplish safe operation in a street environment, the ability
to stop quickly is a must. LRV's are equipped with several
braking systems, including combinations of dynamic, friction
(air) and magnetic (track) brakes, which permits deceleration
of at least 3 miles per hour per second under normal breking,

TROLLEY CAR Ca. 1910

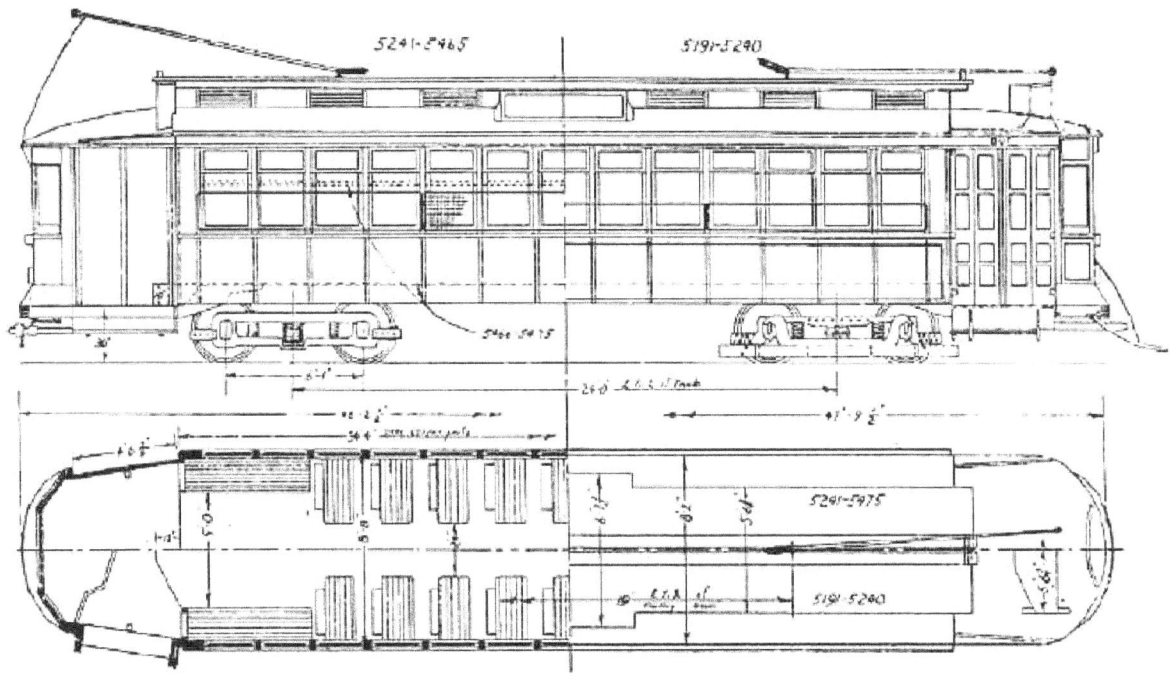

PCC TYPE Ca. 1930 TO PRESENT

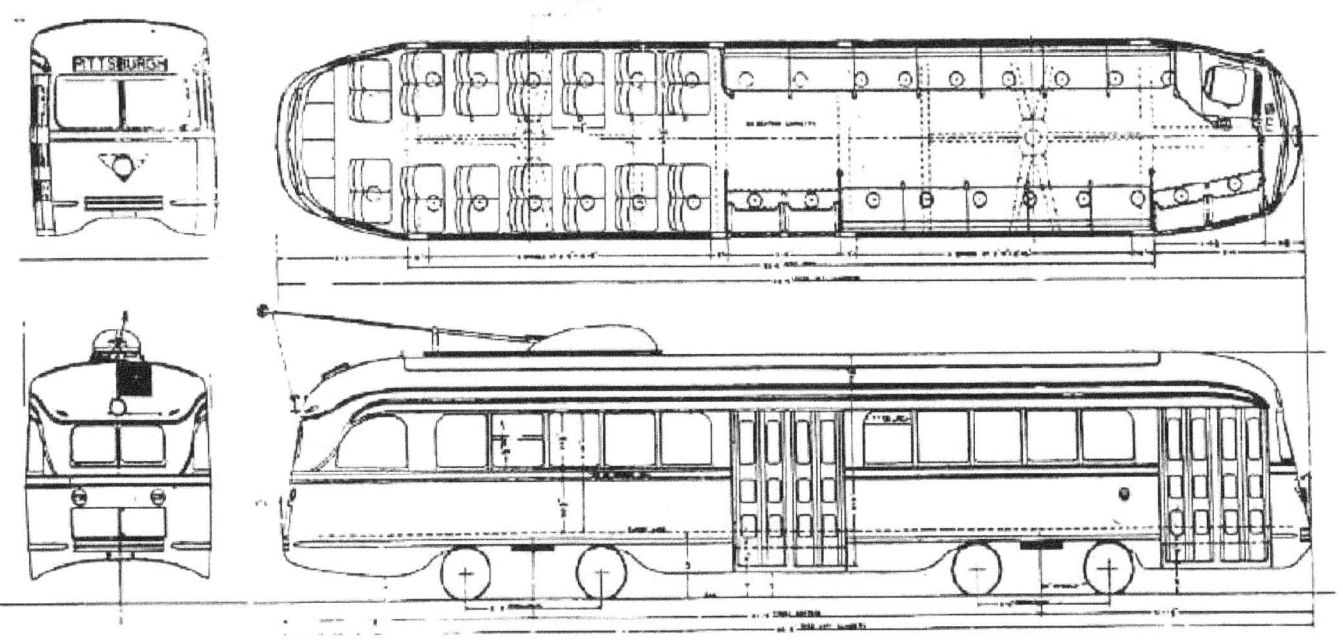

ATTACHMENT D

and up to 6 miles per hour per second in emergencies. At a speed of 12 MPH, an LRV could stop in slightly less than 34 feet using normal service braking, and an emergency stop would require less than 17 feet. From a speed of almost 25 MPH, the braking distances would be 135 feet in normal service braking, and less than 68 feet in emergency braking. Depending on the LRV used, the entrance locations vary from the ends, to a combination of ends and center doors.

Because the trains operate in streets, it is not practical to implement full Automatic Train Operation (ATO). However, there is a lower form of ATO, which would be most useful, as it is desirable for traffic light phasesd and timing to be sensitive to train movements. This type of system is called "Train to Wayside Communications" (TWC). TWC is a communication link system consisting of a carborne transmitter and wayside receiver. The system receives and transmits carborne data for the purpose of dispatching, routing and monitoring by a central control office. The carborne transmitter continuously transmits a frequency shift keyed (FSK) modulated radio frequency carrier signal containing such train data as train stopped, train ready, station check, destination. Train data is inputted to the system by digital thumb wheel switches on the cab console, and various sensors on the train. The FSK modulator then converts the raw data into radio frequency digital bits which are transmitted to a central digital computer where the information is processed and translated into other actions, such as traffic light coordination and train movement updates.

TASK 7

Tracks

For the most part, track would be installed in the center of the roadway. The track would be designed to street railway specifications which would allow vehicular traffic to operate unimpeded. The top-of-rail elevation will be flush with the surrounding asphalt surface to allow the movement of pedestrian and vehicular traffic. The overall width of track construction area for two tracks is 22 feet.

Track Construction

Track construction is a simple procedure: first, pairs of parallel saw cuts, spaced 11 feet apart (or 22 feet apart if double track), and made 18 inches deep, would be cut in the roadway. The area between the saw cuts would then be excavated to a depth of 18 inches. The rails are then temporarily laid

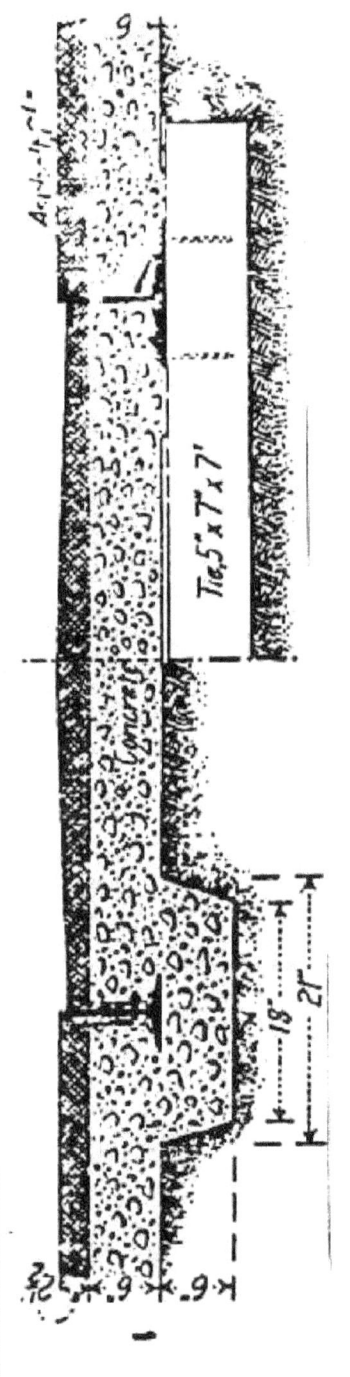

Street Railway Track

ATTACHMENT E

or ties. The 9 inch girder rail (heavy "T" rail may also be used), 80 feet long, are drilled for tie-rods 10 feet apart and for one bolt at each end for temporary splicing. Oak or steel ties, 5 x 7 inches, and 7 feet long, are laid 5 feet apart, alternate ties being tamped with stone ballast, after which the track is surfaced, gauged and lined. The alternate ties are then embedded in concrete, well tamped, and a concrete stringer 6 x 18 inches formed under each rail. All the space between the ties and for 2 feet outside the rails is filled with a bed of concrete, composed of one part Portland cement, 3 parts sand and 5 parts broken stone. This is then allowed to set for 72 hours. The splice bars are then removed and the joints electrically welded. Under this system 2,500 feet of track has been laid in one day. It is unlikely that a contractor will maintain a work force large enough to lay 2,500 feet of track per day due to high labor costs. In New York City, 1,000 feet per week would be reasonable.

TASK 8 *public safety from wire itself*

Power Supply and Distribution *Is there Con Ed capacity in peak load times, how to get power lines to substation*

must show power demand

Power would be transmitted to the LRV's through a single trolley wire, with electrical contact accomplished through the trolley pole mounted on the vehicle's roof. The support system consists of poles mounted along the curb at intervals of 100 feet, which in turn support a single span support wire which holds the electric contact wire. See Attachment "F" for details. Alternatively, where possible, the span wire may be anchored directly to buildings to reduce costs.

There are two possible methods of supplying power to this system. The first would be to tap power directly from existing NYC Transit Authority Substations, a substantial number of which are located in close proximity to the Light Rail line. The Long Island Rail Road also has subdstantial sub-station capacity at the Flatbush Avenue Terminal. Significant modifications to the sub-stations themselves would not be required to provide power for this system, as its power requirements are less than that of a single subway train. Only extra branch switches and circuit breakers need be installed.

The second alternative is much more costly. New York Power Authority power could be supplied through the Con Edison network to two new sub-stations of the 0.750 Megawatt size. Each of these sub-stations would require a main step-down transformer and a rectifier unit for the AC to DC conversion.

ATTACHMENT F
Overhead Wire Concept
For Two Tracks

±17'

±100'

Impacts and Mitigation

Transparent nylon span wire may be used to minimize the visible impact of the overhead wire system. Support poles may be cast to resemble period street lights.

TASK 9

Origination and Termination Points

The primary origination and termination points on this system are Brooklyn Pier 6 and Flatbush Avenue/LIRR. The origination/termination points of the feeder branches are the foot of Van Brunt Street in Red Hook, and the intersection of Vanderbilt Avenue and Plaza Street in Park Slope. Terminal train turning is accomplished by a simple set of track switches which will permit the vehicles to cross over to the other track, and reverse direction. Reverse loops would also be provided at Hanson Place and Cadman Plaza East.

TASK 9A

Passenger Boarding

Passenger boarding will essentially be from street level. The terminals on the main line will see the highest passenger service levels.

TASK 9B

Employee Distribution

Trains will be controlled and monitored by a central control office at the Pier 6 terminal. Therefore, the only employees on the road will be the actual train operators. We anticipate a total of 22 employees, 18 of which will be train operators on the road. The remaining 4 will be administrative and repair personnel based at the Pier 6 terminal.

TASK 10

Parking

No additional parking is projected for the accomodation of this project. Most LIRR stations on Long Island have sufficient Park and Ride facilities. Persons living in the downtown area who would use this system would simply leave their automobiles at home. It is not foreseen that people will drive in from other areas, and then use this system.

TASK 11

Fare

It is a primary goal to keep the fare as low as possible in order to compete with other transportation modes, and attract people away from their automobiles. To gain some perspective on price structuring, express buses charge $3.00 per one-way trip. A private ferry operation began between Sheepshead Bay and Wall Street. They charge $9.00 per round trip.

Based on preliminary calculations, our fare would be $5.00 per round trip, including the ferry ride. Transit Cheks would also be honored. Magnetic swipe tickets would be used. Railroad clerks would not be needed. Tickets would be purchased from vending machines at convenient locations. Monthly commuter ticket passes would also be available. Expired tickets would be confiscated by the carborne ticket reading machine.

TASK 12

Train Storage

The tunnel portals would be fitted with iron bar gates as they were originally. The trains could easily be laid-up inside the tunnel when the line is not in use. Further, the tunnel would be equipped with electronic sensors wired directly to the local police precinct.

TASK 13

Train Repair and Maintenance

Pier 5 would be the ideal location for a small repair shop.

It would be able to accomodate four cars. It would contain a platform for car cleaning. Typical equipment would include four 25-ton car jacks, car supports, truck turntables and a 10-ton monorail crane. Other equipment would include both electric and gas welding sets, various power tools, electric lathes, lubricants and cleaning solvents. Environmental safeguards commensurate with similar NYCTA shops would be implemented.

TASK 14

Tunnel to Street Surface Interface

The original western tunnel portal is located just west of the
BQE north bound Atlantic Avenue exit ramp. It is in the center
of Atlantic Avenue. The bottom of the portal is about 21 feet
below the grade. There is a stone approach ramp at the mouth
of the portal which brought the track to grade just east of the
intersection of Columbia Street and Atlantic Avenue. See
Attachment "G" for details of the proposed reopening. The east
portal is located midway between Boerum Place and Court Street
in the center of Atlantic Avenue. The concept is the same as
indicated in Attachment "G". The present number of traffic
lanes will be maintained by narrowing the sidewalk around the
portals and approach ramps. Refer to drawing for details.

TASK 15

Traffic Considerations

No streets are removed from service, and around the block
circulation is very much as present. The route alignment is
designed with special consideration given to certain key
factors. Specifically :

FIGURE 29
DOWNTOWN BROOKLYN
GENERALIZED TRAFFIC FLOW
8AM – 9AM

BASED ON NYC DEPARTMENT OF TRANSPORTATION COUNTS

ALTERNATIVE ALIGNMENTS

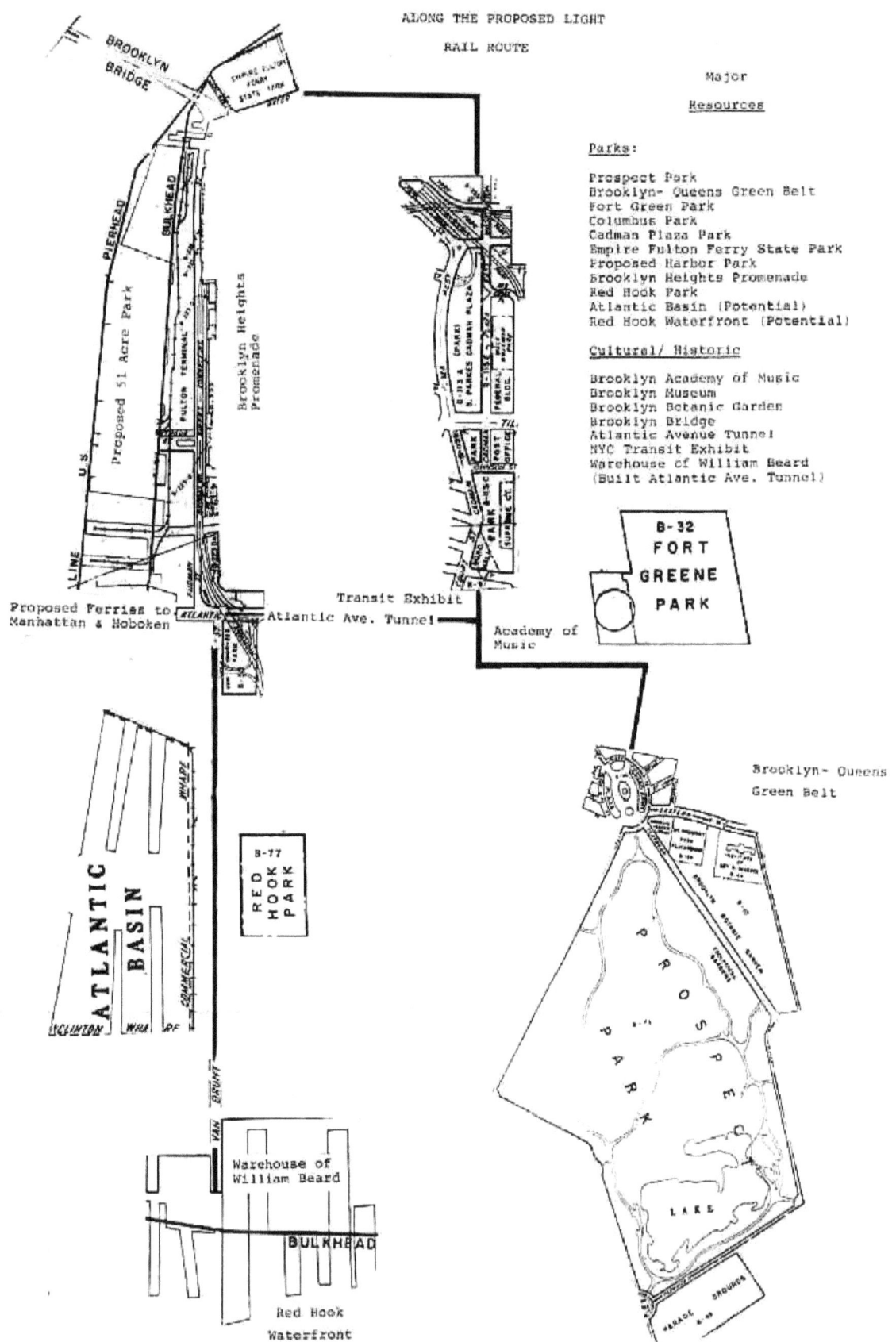

ATTACHMENT G

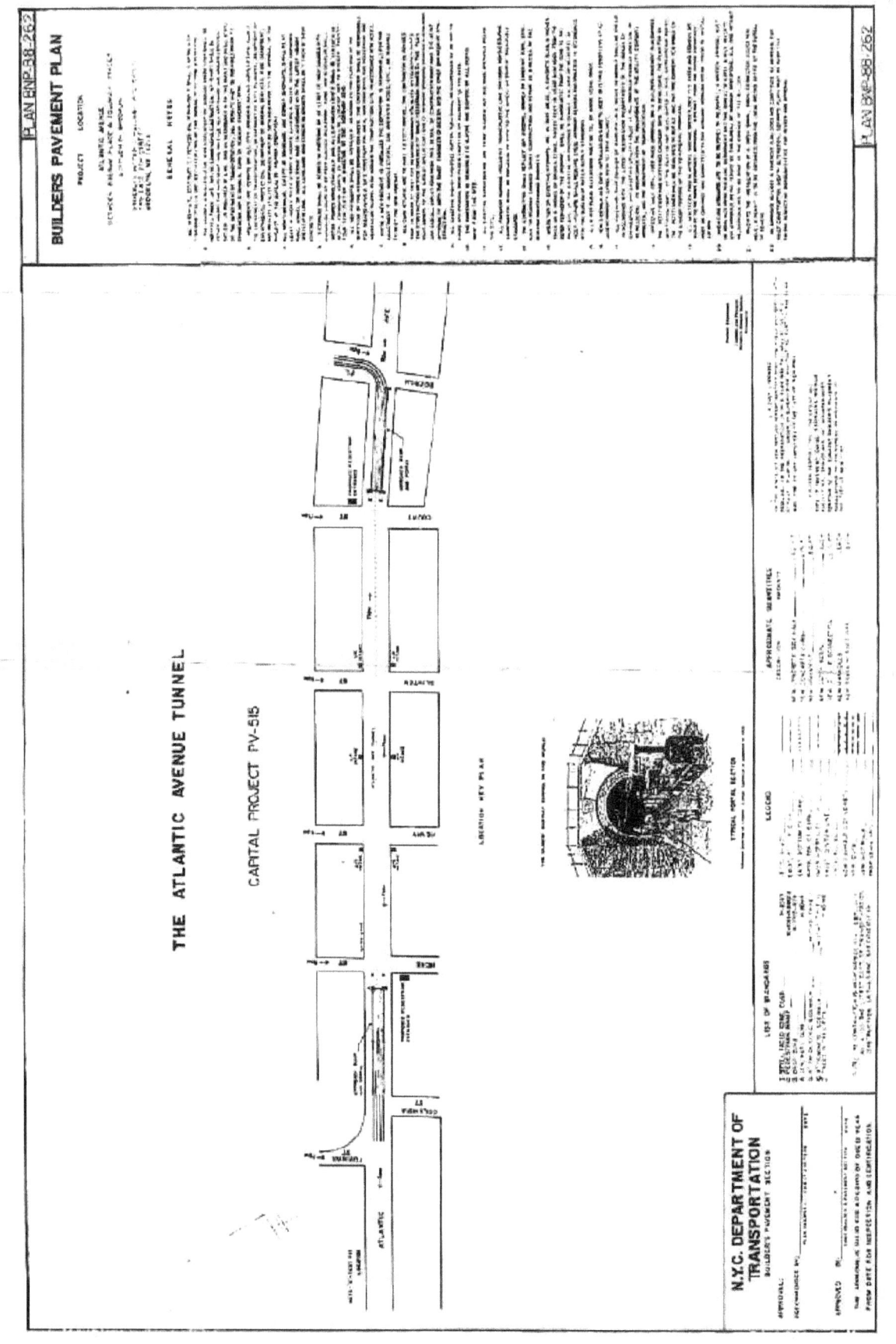

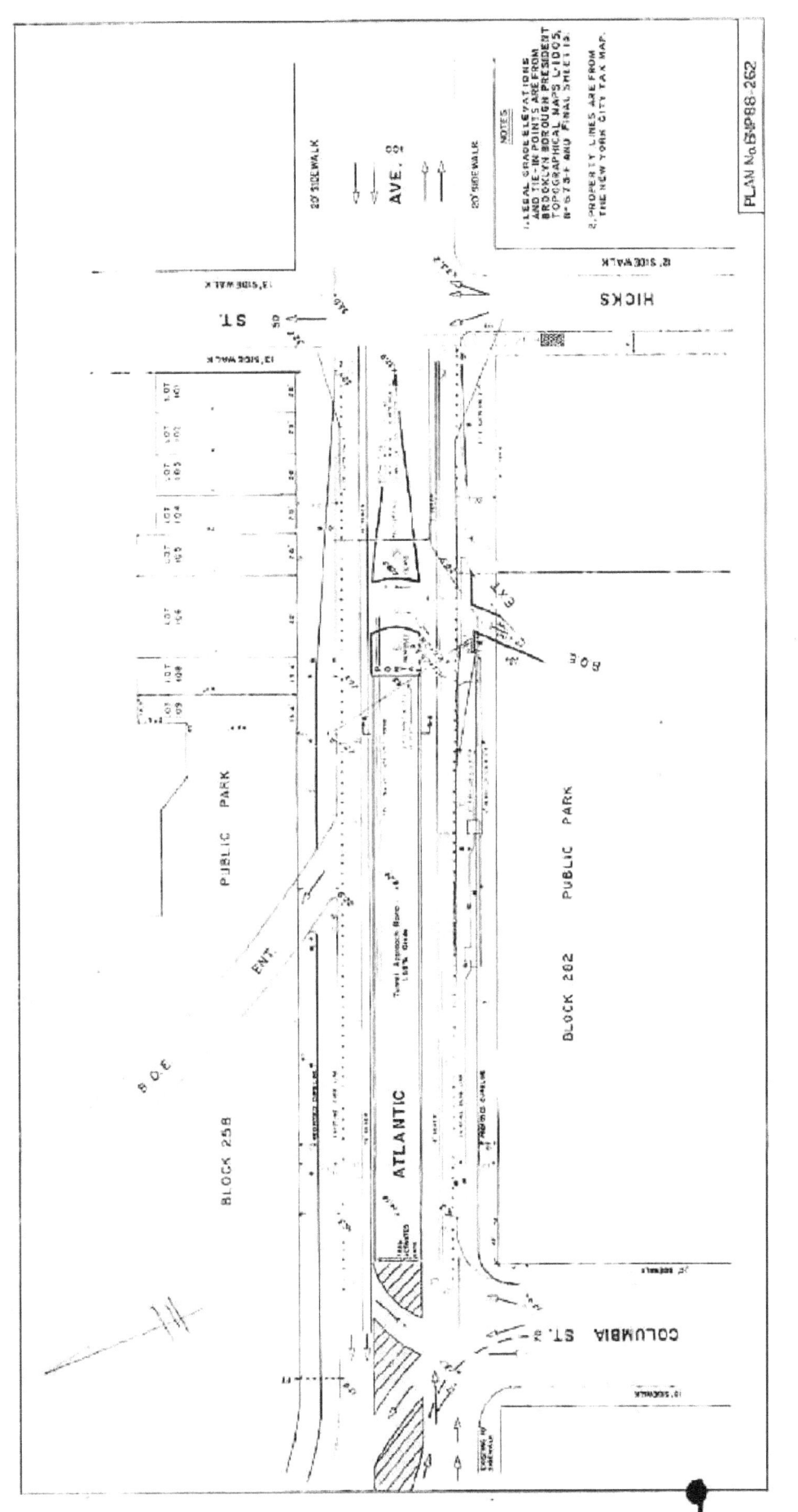

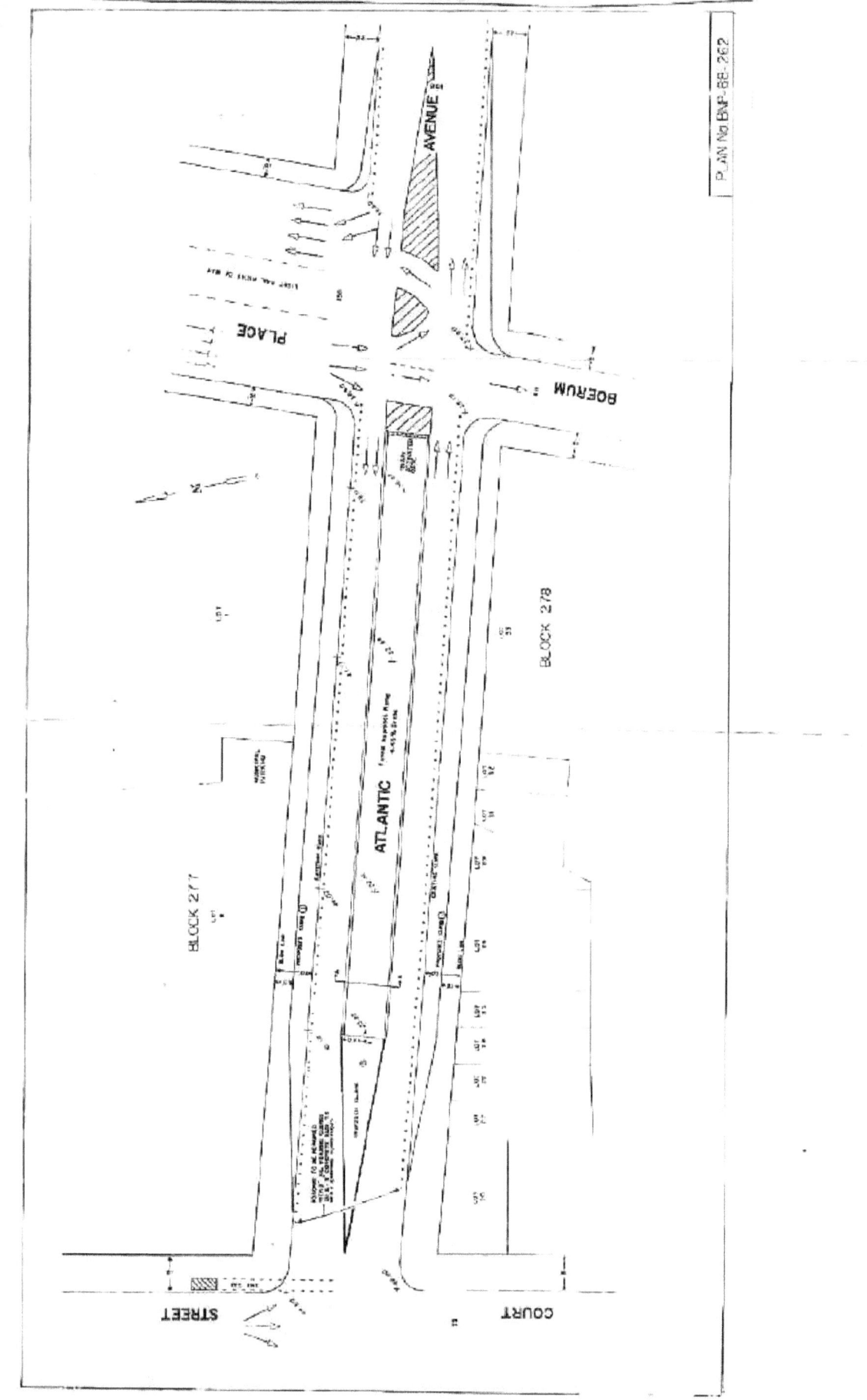

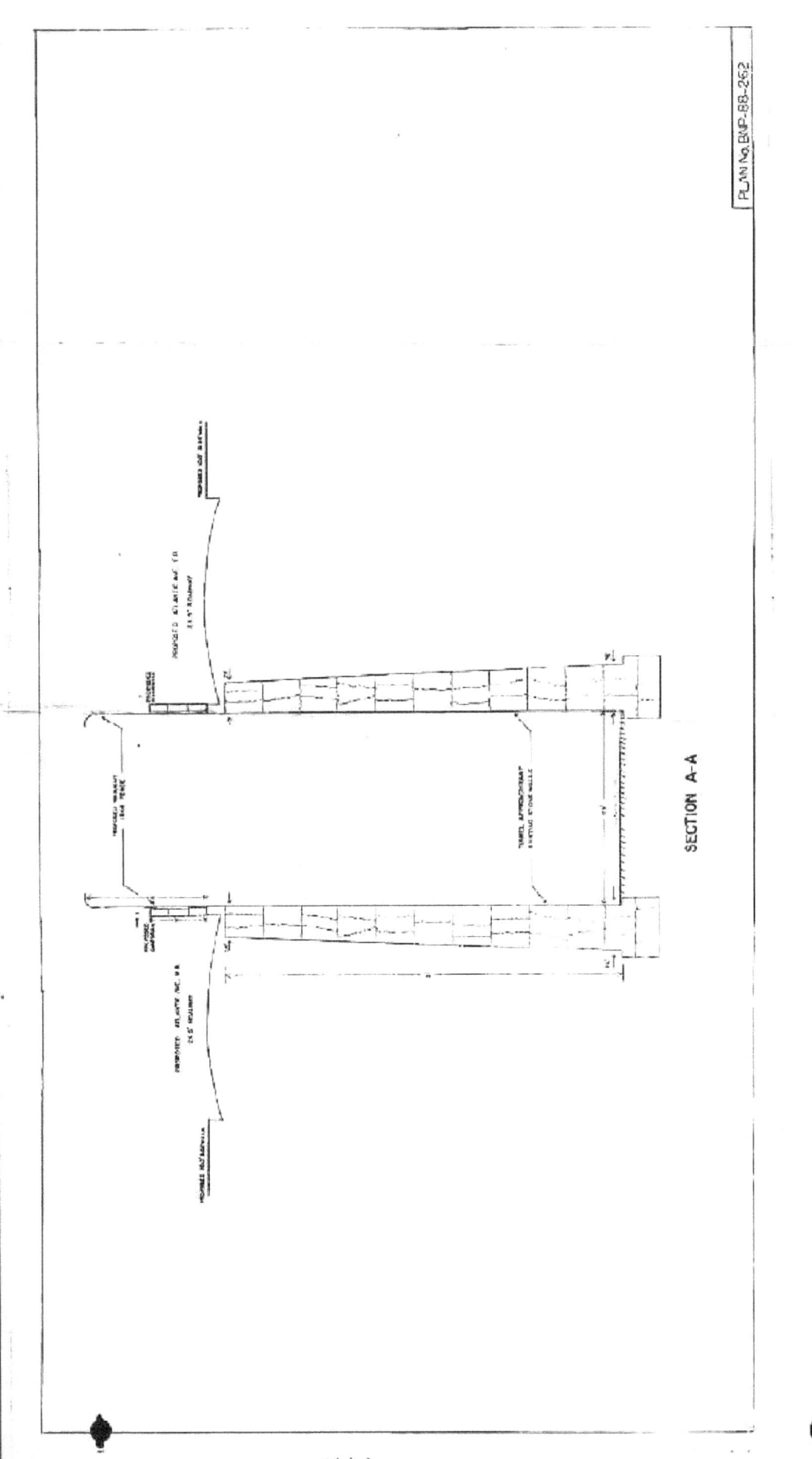

a) Limit left and right hand turns;

b) Use streets over 30 feet in width;

c) Provide no turns onto or off of Flatbush Avenue;

d) A minimum 10-foot travel lane for general traffic around tracks and loading areas;

e) Single track operation along feeder routes, where required;

f) Full protection of train turns from conflicting through or turning traffic. Train actuated signal phases would operate within overall background cycles at key conflict points.

g) Since an LRV has much greater acceleration than a Bus, right turn time for an LRV will be equal to or less than, that of a Bus.

Impacts and Mitigation

LRV movements will have to be blended into overall background signal phase cycles. Additional parking regulations would be implemented. From a psychological viewpoint, the public would have to become accustomed to the presence of LRV's.

TASK 16

Noise

Light Rail vehicles operating on well maintained track should be significantly quieter than diesel buses. LRV interior noise levels should range from 67 to 70 decibels, while bus interior noise levels range from 70 to as much as 85 decibels (during acceleration). Exterior noise levels, which measure the impact on the surrounding community, are in the range of 75 decibels for LRV's, while bus noise levels range from 80 to almost 90 decibels on acceleration.

Impacts:

Since LRV's are quieter than buses, ambient noise levels along the routes will be reduced.

Conceptual agreements between me & parks dept.

TASK 17

Open Spaces Recreational Resources

The proposed Light Rail routes will interconnect all major downtown Brooklyn parks and cultural facilities with eachother; Brooklyn communities, Manhattan and New Jersey (via ferry connection), as well as provide easy pedestrian and bicycle access to the Brooklyn-Queens Greenbelt. Some of the LRV's would be provided with bicycle racks. See schematic drawing.

This would improve public access to public parks and recreational resources, thereby increasing their patronage.

What activities excluded, safety of patrons, current activity use

Impacts and Mitigation:

The Light Rail route would require a 10-foot wide right-of-way through Columbus Park. See diagram. The proposed alignment would only require the relocation of one small tree. Safety impacts would be mitigated by restricting LRV speed through the park, and by separating the right-of-way by a decorative cast iron fence. Automatic overhead wire shut-off would be provided in case of contact wire support failure.

TASK 18

Air Quality

The proposed Light Rail System would reduce air pollution in downtown Brooklyn by reducing the number of automobile and bus trips into the area. For example, the new Light Rail System in San Diego has attracted 47% of the passengers who were using five bus routes parallel to the Light Rail. Source: Assessment of the San Diego Light Rail System, November, 1983: Further, it has been demonstrated that Light Rail
*See Table 5.4-6.
Systems attract atleast 30% more passengers than bus routes. This represents the attraction of people who presently use automobiles.
Source: Transportation and Land Use Policy by Boris Pushkarev, pp18.
*See Appendix G, Regional Plan Assn. Book by Boris S. Pushkarev & Jeffrey M. Zupan, 1977
In downtown Brooklyn, the Light Rail attraction rate would be still higher, due to the 'amusement park' factor of the historic tunnel attraction, and vintage trolley cars on the system. The DEIS will give a detailed passenger analysis which will indicate the total number of trips which will be diverted from fossil fueled vehicles. For example:

TABLE 5.4-6: MONTHLY REVENUE PASSENGER COMPARISON
SDTC SOUTH BAY CORRIDOR
BUS AND TROLLEY
FOUR-MONTH PERIOD, OCTOBER-JANUARY

	OCTOBER		NOVEMBER		DECEMBER		JANUARY	
	1980	1981	1980	1981	1980	1981	1980	1981
5 Bus Routes	439,966	219,622	438,769	203,014	450,110	207,634	442,508	197,132
Trolley*	-	301,291	-	297,907	-	298,861	-	300,582
Total	439,966	520,913	438,769	500,921	450,110	506,495	442,508	497,714
% Monthly Increase/Decrease		+18.4%		+14.2%		+12.5%		+12.5%

* Trolley revenue passengers include Vendomat Ticket Purchases plus Regular Pass riders (see Appendix B calculation). Not included in these estimates are illegal riders, transfers, or elderly and handicapped passengers.

Source: Reference (10).

The EIS for MetroTech describes the number of trips by cars travelling to and from their site per day from Long Island and New Jersey as follows:

2797.8 person trips per day from Long Island by Car
 607.7 person trips per day from New Jersey by Car

3405.5 Total from Long Island and New Jersey

Note: see Appendix "C" for calculations and documentation.

Assuming a 250-day per year work year, this yields 851,375 person trips per year by auto from Long Island and New Jersey alone. As outlined earlier, subway connections between the LIRR terminal and MetroTech will be overburdened, as per the projections in the MetroTech EIS. Light Rail would provide comfortable and speedy access from the LIRR terminal to MetroTech which would provide a mass transit alternative via the LIRR to driving in from Long Island by car. Hoboken is a major terminus of New Jersey public transportation. A ferry connection from Hoboken to the Brooklyn Light Rail link to MetroTech would capture a certain segment of those travelling from New Jersey to MetroTech by car. If a fifty percent share of these segments were captured by Light Rail, this would eliminate 425,687.5 automobile person trips per year into downtown Brooklyn, with the attendant improvements in air quality and traffic congestion.

As indicated, this is only one example of a particular Light Rail market segment. The EIS will address all segments with detailed analysis - in particular, local bus and automobile trips which would be eliminated.

TASK 19

Traffic Impact Analysis

The Downtown Brooklyn LRT Line will either operate on exclusive rights-of-way or be carefully integrated into the overall street pattern. No streets are removed from service, and "around the block circulation" is very much as at present. A few additional curb parking restrictions are required, but some of the restrictions apply only to the peak hours.

This section describes specific roadway and intersection treatments to show how the LRT would interface with traffic and pedestrians at critical points along the line. These treatments define paths of travel for each type of vehicle:

signal control requirements; and steps to be taken to maintain sufficient traffic capacity, movement and flow.

Fulton/Ft. Greene/Lafayette (Figure 2) - Fulton Street is widened 5-feet on the north side to provide a 46-foot wide road. The widened cross section includes two westbound travel lanes (A.M. peak), totalling 20 feet; a 10-foot eastbound bus and LRT lane, a 5-foot passenger loading island, and an 11-foot eastbound traffic lane. The eastbound Fulton Street traffic is about 400-500 vehicles per hour and can be accommodated in a single lane.

If the widening along the north curb of Fulton Street adjacent to the transit stop is not practical, then the north curb should be kept at its present location. This will provide a 10-foot travel lane and a 7-foot parking lane. In both cases, however, parking should be prohibited during the morning peak-hour along this curb. There should be no parking at any time along the south curb.

The LRT line shares the center lane of Ft. Greene Place with general traffic flow. Parking remains along both curbs to provide maximum convenience to residents. LRT operation in mixed traffic poses no problem, since traffic flows along Ft. Greene Place are very light - about 50 vehicles in the peak hour.

<u>Hanson Place/Fort Greene/Ashland</u> (Figure 3) - Hanson Place is widened along the south curb to provide a 45-foot wide roadway.

Two westbound LRT tracks are provided in the northern half of the street, one being adjacent to the curb to facilitate passenger access. A 5-foot loading platform is also provided to the south of the second track. The eastbound traffic, about 300 vehicles in each peak hour, would operate in two 10-foot lanes in the southerly part of Hanson Place.

Between St. Felix and Ashland Place, a single westbound track is provided in the center of the north half of Hanson Place. This permits a 10-foot westbound traffic lane adjacent to the north curb for property access. The radius in the northeast corner of the Hanson-Ashland intersection is increased to accommodate LRV turns.

Traffic signal timing at the Flatbush/4th Avenue intersection is revised to incorporate a "LRV actuated phase". During this special phase, the northbound traffic at Ashland Place is required to stop to permit LRV's to turn right into Ashland Place without conflicts.

The plan assumes that Flatbush and Atlantic Avenues would be widened in this area as recommended in the <u>Downtown Brooklyn Traffic and Transit Study</u>.

Ashland Place/Hanson/Fulton (Figure 4) - The LRV's operate in the center northbound lane of Ashland Place from Hanson to Fulton, sharing a lane with northbound traffic. They operate two-way in the center of Fulton Street west of Ashland in "transit only" lanes. There is a northbound stop along Ashland on the near side of Lafayette Street, and an eastbound stop along Fulton on the near side of Ashland. A 5-foot raised passenger island is provided alongside each stop.

The plan incorporates several street widenings that are needed to provide space for loading islands and an adequate number of lanes for peak-hour traffic. (1) Ashland Place is widened from 43 to 48 feet south of Lafayette to provide 4-full lanes plus a 5-foot island, (2) Ashland is widened from 30 to 40 feet north of Fulton to provide two lanes each way. This widening was recommended in the Downtown Brooklyn Traffic and Transit Study to reduce peak hour queues, and permit reversions in signal sequences. (3) Fulton Street is widened from 42 to 48 west of Ashland to provide a 5' transit stop.

A three phase signal operation separates conflicting LRV and traffic movements and also accommodates the heavy south-to-east left turns from Ashland to Fulton. This phasing is as follows:

-Phase A : Fulton St. E-W and LRV EB

-Phase B : Ashland NB and LRV NB

-Phase C : Ashland SB and WLRT

Existing curb parking restrictions are retained along Ashland Place. Parking is prohibited along both sides of Fulton west of Ashland. The alternative is to let cars use transit lanes midday and to only restrict parking during peak periods.

At Hanson Place and Fort Greene Place, part of the LRT line continues east along Hanson Place to Fulton Street.

Fulton/Boerum/Joralemon (Figure 5) - The LRT route leaves the Fulton Mall at Adams and proceeds northerly through the park right of way to Cadman Plaza east at Johnson Street. Passenger stops are provided along the Park right of way.

Crosswalks are channelized and protected to assure safe pedestrian crossings of both street and the LRT right of way. The signal phasing is similar to the existing phasing; however, an additional LRT-actuated phase has been introduced within the overall cycle. During this phase, pedestrian crossings of the LRT line in the northwest quadrant of the intersection would be prohibited.

The LRT line through the park roughly follows the historic trolley right of way. A crossover is provided between Johnson and Fulton to improve operating flexibility.

At the intersection of Boerum/Fulton/Joralemon, part of the LRT line branches off and proceeds south along Boerum Place. The LRT line occupies a dedicated left lane in both directions. This would leave two 11-foot and two 10-foot traffic lanes in each direction. The existing traffic island in Boerum Place would be narrowed to 4 feet in width. The island would be used for passenger loading at Livingston Street. No stopping would be permitted on Boerum Place during peak hours.

Cadman Plaza East - Johnson Street to Tillary Street (Figure 6) - The LRT line leaves its own right of way at Johnson Street and proceeds northerly in the center of Cadman Plaza East. 5' wide passenger loading platforms are provided adjacent to the old Post Office facilities. A third track for train turnback (short turn operation) is provided immediately north of this stop - the precise configuration will depend upon the length of block between Johnson and Tillary:

-Assuming adequate length, a crossover would be provided to the north of the passenger area preferably with a third track for car layover.

-If this is not practical, then a counterclockwise loop via Cadman Plaza East (NB), Tillary (WB), Cadman Plaza West (SB), and Johnson (EB) should be considered.

The plan, as shown, permits two phase traffic signal operations along Cadman Plaza East at Johnson and Tillary Streets. Auto left turns at the Tillary-Cadman Plaza East intersection are prohibited to minimize car-LRT conflicts.

Washington at Prospect (Figure 7) - The LRV's operate two-way in the center of Washington Street in mixed traffic. Curb parking is prohibited (as at present) along Washington Street. The sidewalks are narrowed to 5 feet at stops to provide a 3' painted passenger island plus a 10-foot travel lane each way at stops.

Waterside - The LRT line is closely coordinated with the proposed Fulton Landing development. It follows the existing rail alignment along Plymouth Street. The line, double-tracked throughout, has stops (1) between Main and Washington Streets near the Sports and Recreation center and (2) at the foot of New Dock Street alongside the State Park and (3) under the Brooklyn Bridge near Fulton Landing.

New Dock/Water (Figure 7.1) - New Dock Street is a dead-end street with no vehicular traffic. It can be used exclusively by the LRT. Water Street would be made two-way from Main Street to Fulton Street. No standing any time would be permitted on Water Street between New Dock Street and the River Cafe.

The widened Ashland Place north of Fulton Street as recommended in the <u>Downtown Brooklyn Traffic and Transit Study</u> will permit a three-phase signal operation. Together they will alleviate a currently congested intersection. Adjustments in signal timing and sequences at other locations will allow LRV's to move safely, without eroding essential street capacity.

<u>Atlantic (Boerum/Columbia)</u> - (Figure 8) - At Boerum Place the LRT line turns west and enters a ramp leading into the Atlantic Avenue Tunnel. Passenger loading would be accomplished in the ramps. The roadway in Atlantic Avenue from Boerum Place to Court Street and from Hicks Street to Columbia Street would be widened to 80 feet. No stopping anytime would be permitted along these segments.

At Columbia Street, the LRT line comes to grade and proceeds along Atlantic Avenue to the Pierhead line where the Ferry Terminal would be located.

At the intersection of Atlantic Avenue, part of the LRT line branches off and proceeds south along Columbia Street.

Columbia Street

The roadway in Columbia Street varies in width from 50 feet near Atlantic avenue to 36 feet near Woodhull Street. In areas of Columbia Street where parking is permitted, there is one moving lane in each direction. The B-61 bus presently operates along this part of Columbia Street. Bearing in mind potential future activity along Columbia Street, the following would be preferred modifications on Columbia Street for Light Rail purposes:

-Widen the narrower parts of the roadway in Columbia Street to 40 feet as part of HWK 700BW. Widen the roadway in Columbia Street to 60 feet between Atlantic Avenue and Amity Street

-Widen the roadway in Columbia Street by an additional 5 feet adjacent to LRT loading islands, as part of HWK 700BW.

-Impose No Standing regulations along Columbia Street from Atlantic Avenue to Carrol Street during peak hours.
-Close the BQE entrance on Columbia Street (nr. Atlantic Avenue) between the hours of 3PM to 7PM to improve traffic flow
This would permit a 10-foot LRT lane and a 10-foot moving lane in each direction, as well as provide space for 5-foot passenger loading islands.

Columbia/President/Carrol (Figure 9) :

At these intersections the LRT line splits. The southbound track turns West along President Street, and the northbound track West along Carrol Street. The tracks then reconverge at Van Brunt Street & Carrol Street. LRT operation would be west on President Street and East on Carrol Street. The roadway in

President and Carrol Streets is 35 feet wide. This allows ample space for a 15-foot moving lane/LRT lane and two 10-foot parking lanes.

Van Brunt Street

Van Brunt Street has a 36-foot wide roadway. There is one moving lane of traffic in each direction. The B-61 Bus presently operates along this part of Van Brunt Street. Bearing in mind potential future activity along Van Brunt Street, the following would be preferred modifications on Van Brunt Street for Light Rail purposes:

-Widen the roadway to 40 feet as part of HWK-700A

-Widen the roadway an additional 5 feet along Van Brunt Street adjacent to LRT loading islands, as part of HWK-700A

-Impose No Standing Anytime restrictions along Van Brunt Street from President Street to the Pierhead Line during peak hours.

This would permit a 10-foot LRT lane and a 10-foot moving lane in each direction, as well as provide for 5-foot passenger loading islands.

Hanson Place

Both LRT tracks would be positioned on the north side of Hanson Place. Parking along the north side of Hanson Place from Ashland Place to Fulton Street would not be permitted. The roadway in Hanson Place is 48 feet wide, which will allow space for two LRT tracks and two 10-foot moving lanes.

Fulton/Hanson (Figure 10):

At this intersection the LRT route would turn onto Fulton street. At this point Fulton Street's roadway is 49 feet wide. It is proposed that the roadway be widened to 60 feet between Greene Avenue and Vanderbilt Avenue. This could be accomplished during the planned reconstruction of Fulton Street. No Standing during peak hours would be permitted on Fulton Street between Hanson Place and Vanderbilt Avenue.

Fulton/Vanderbilt (Figure 11):

At this intersection, Fulton Street and Vanderbilt Avenue both have 44-foot roadways. It is proposed that the roadway in Vanderbilt Avenue be widened to 60 feet as part of the rebuilding of Vanderbilt Avenue (HWK-715). No Standing during peak hours would be permitted on Vanderbilt Avenue between Fulton Street and Atlantic Avenue.

Vanderbilt Avenue:

Vanderbilt Avenue widens to a roadway width of 70 feet. This would permit a 10-foot LRT lane, south of Atlantic Avenue, a 10-foot moving lane, and a 10-foot parking lane in each direction, as well as space for a 10-foot wide median island for passenger loading. See Figure 12 for a diagram of the LRT terminal at Vanderbilt Avenue & Plaza Street.

Piers 1-6:

The LRT would either operate on a right-of-way through the proposed Harbor Park, or on the west side of Furman Street. Proposals to widen Furman Street have been put forth by other groups. If this came to pass, the LRT would operate in the widened part of said street. The alignment through Harbor Park would be preferred.

TASK 20

Public Transportation:

No Build Scenario, 1993:

Downtown Brooklyn is a nexus of several arterial subway lines and bus routes. However, the many downtown Brooklyn

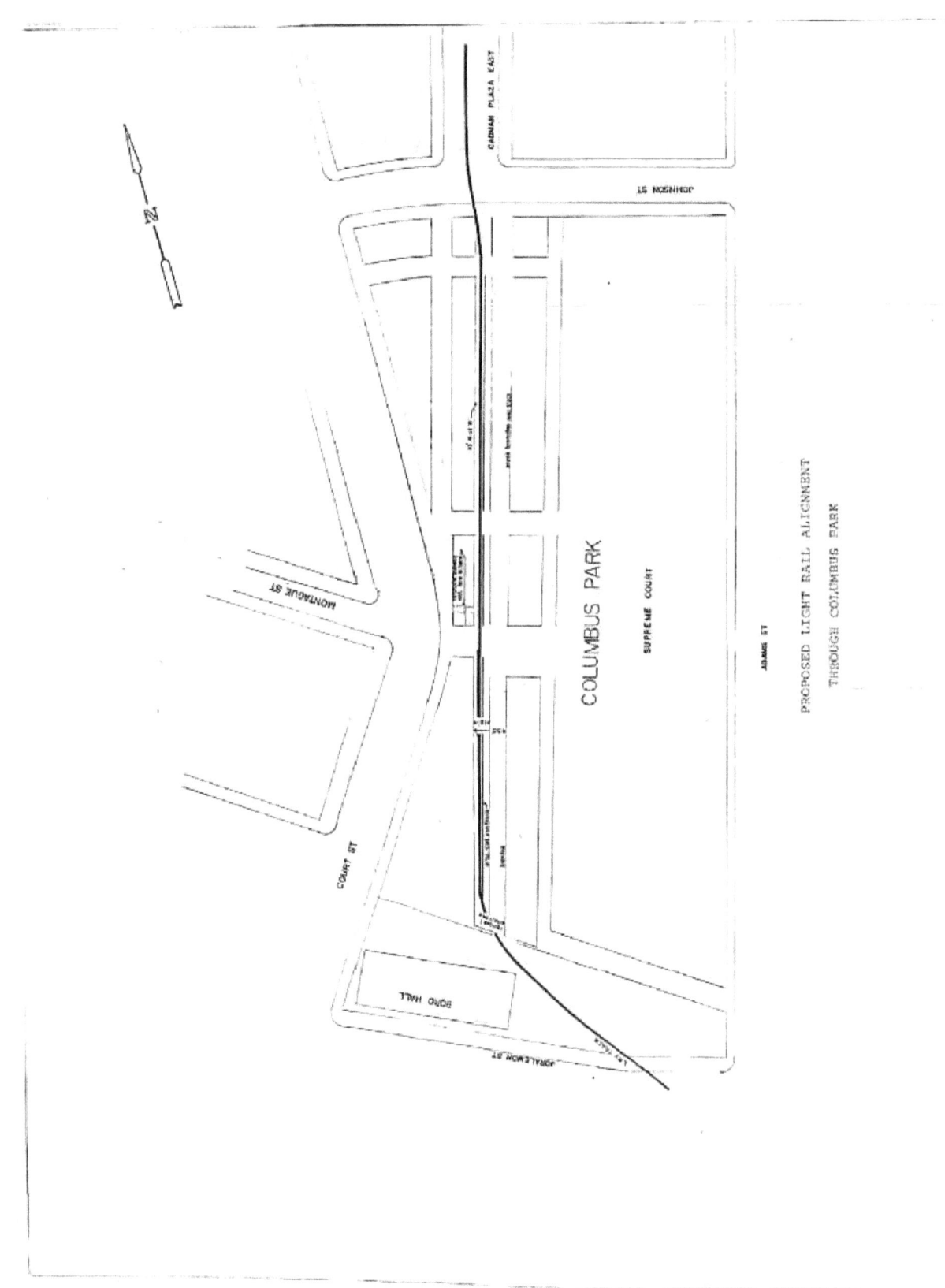

PROPOSED LIGHT RAIL ALIGNMENT
THROUGH COLUMBUS PARK

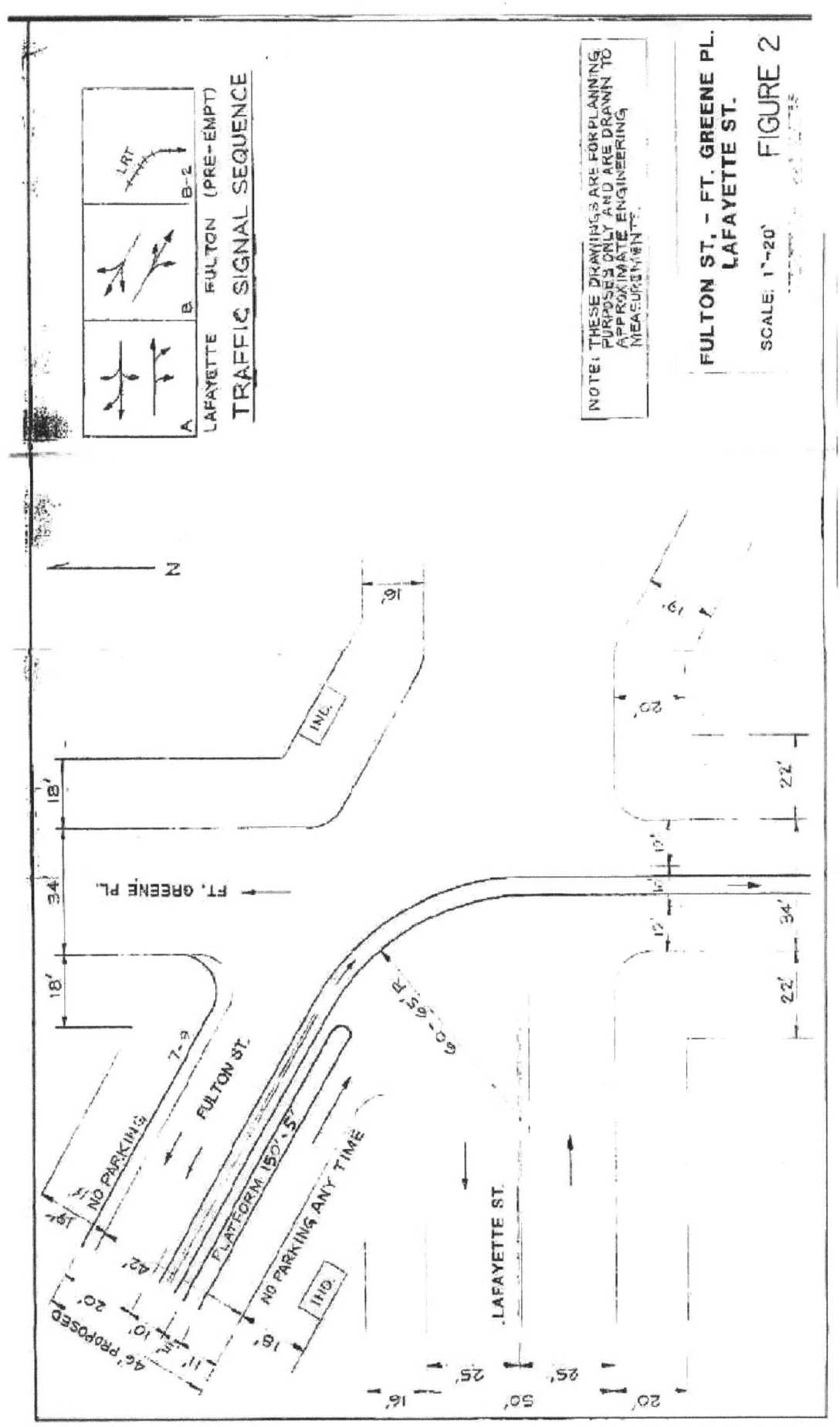

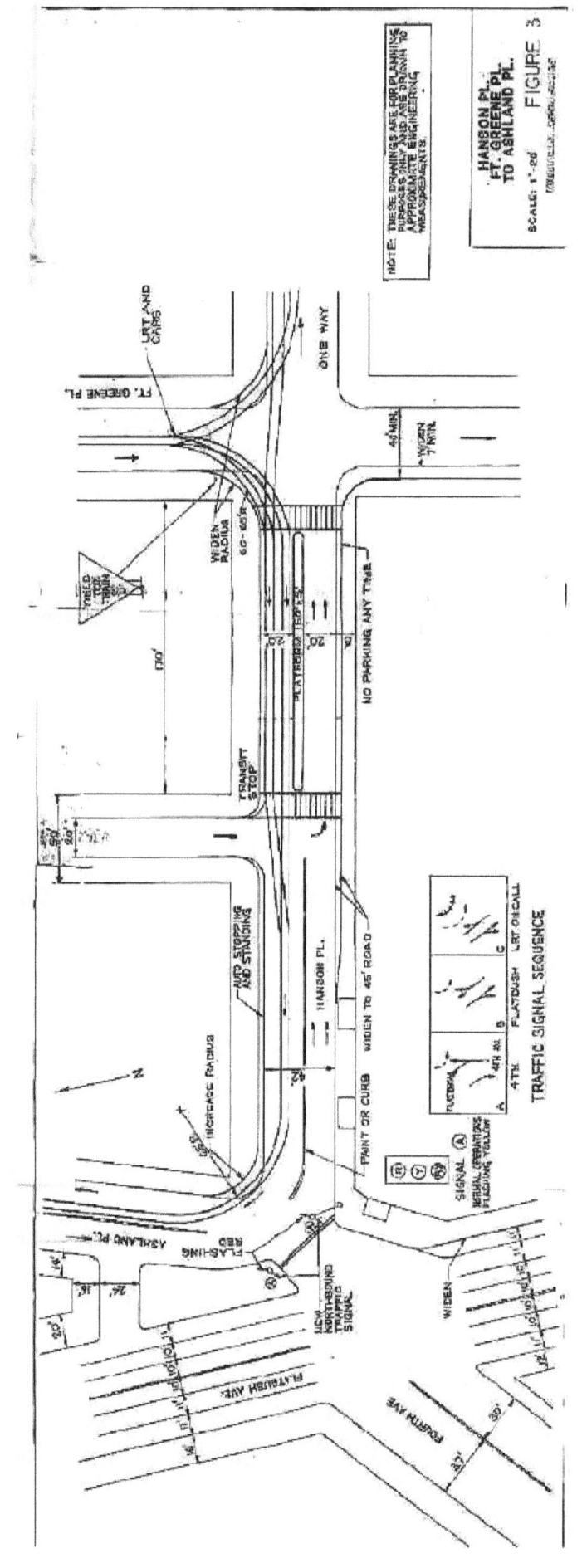

HANSON PL.
FT. GREENE PL.
TO ASHLAND PL.
SCALE: 1"=20' FIGURE 3

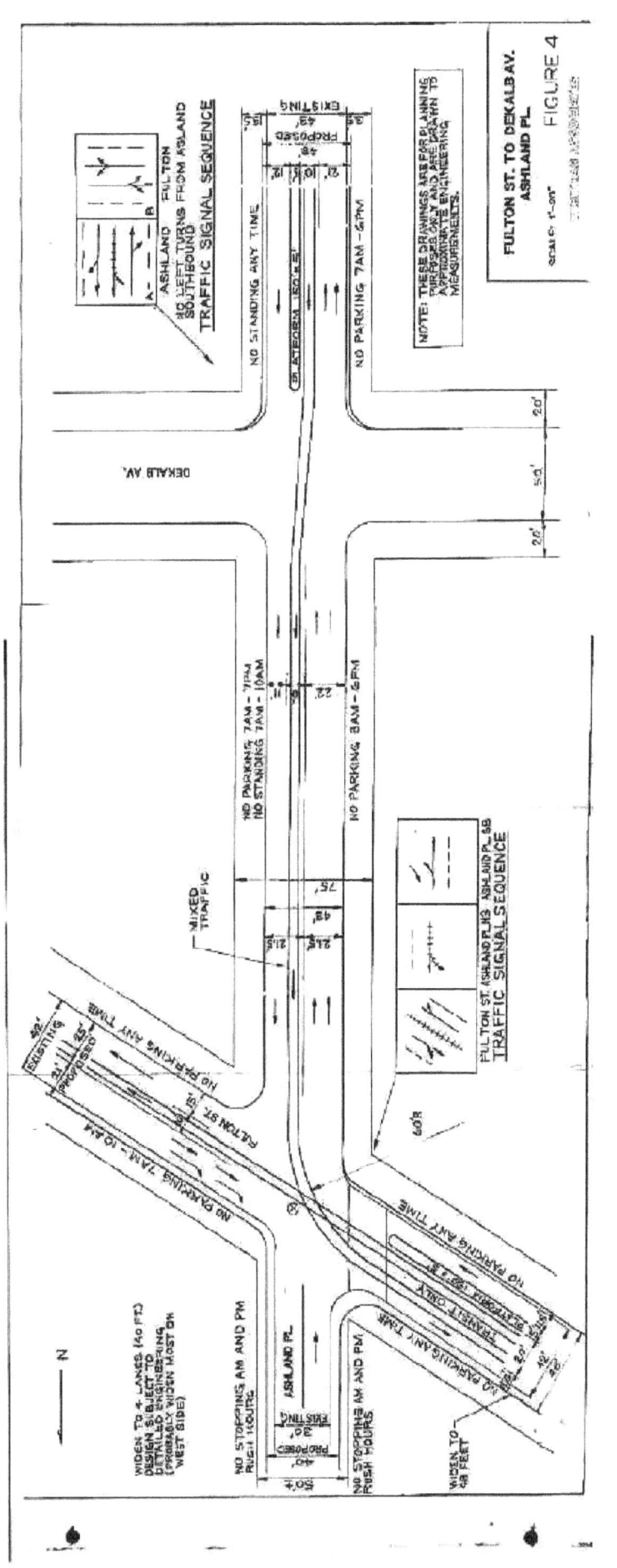

FULTON ST. TO DEKALB AV.
ASHLAND PL.
FIGURE 4

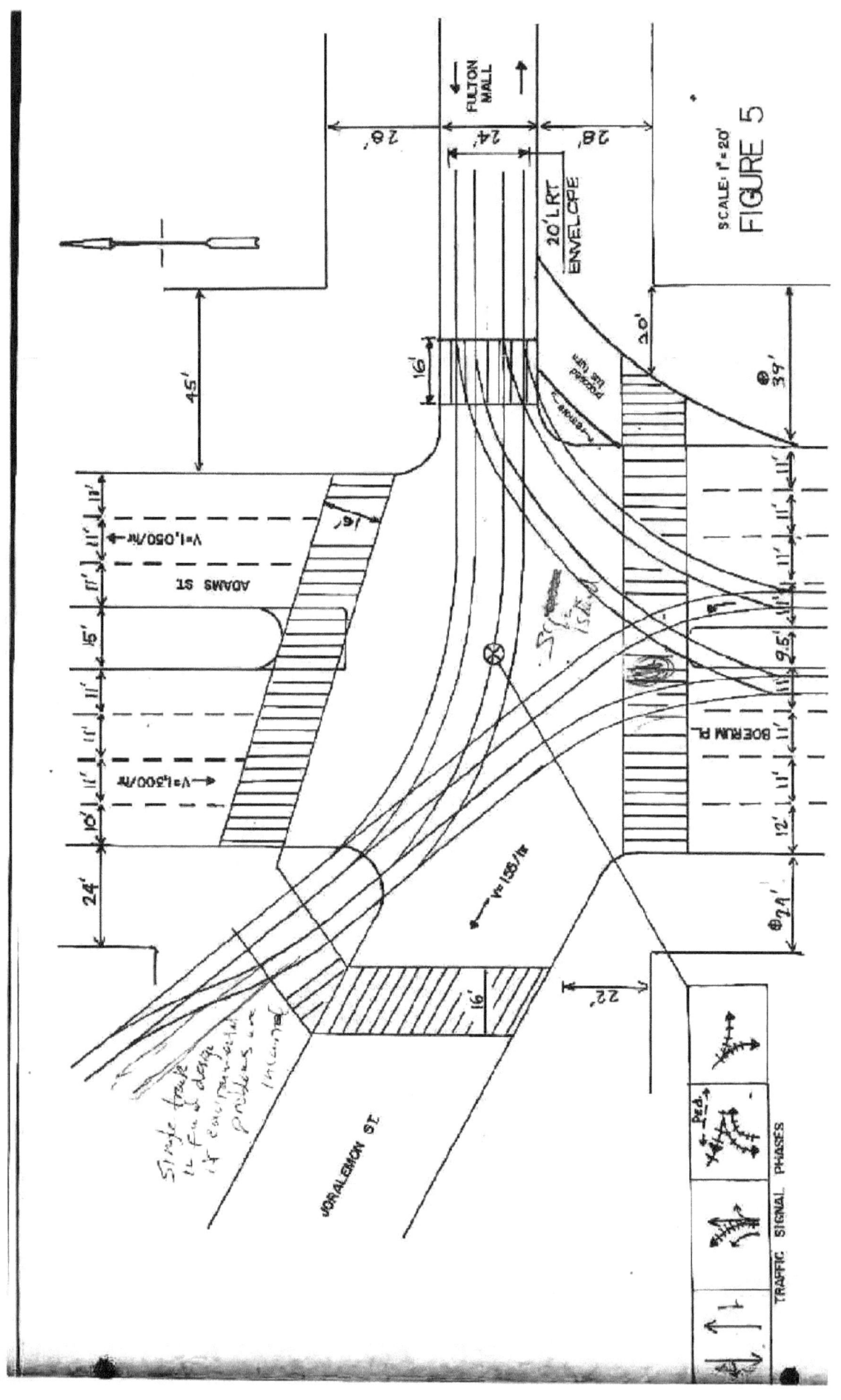

FIGURE 5

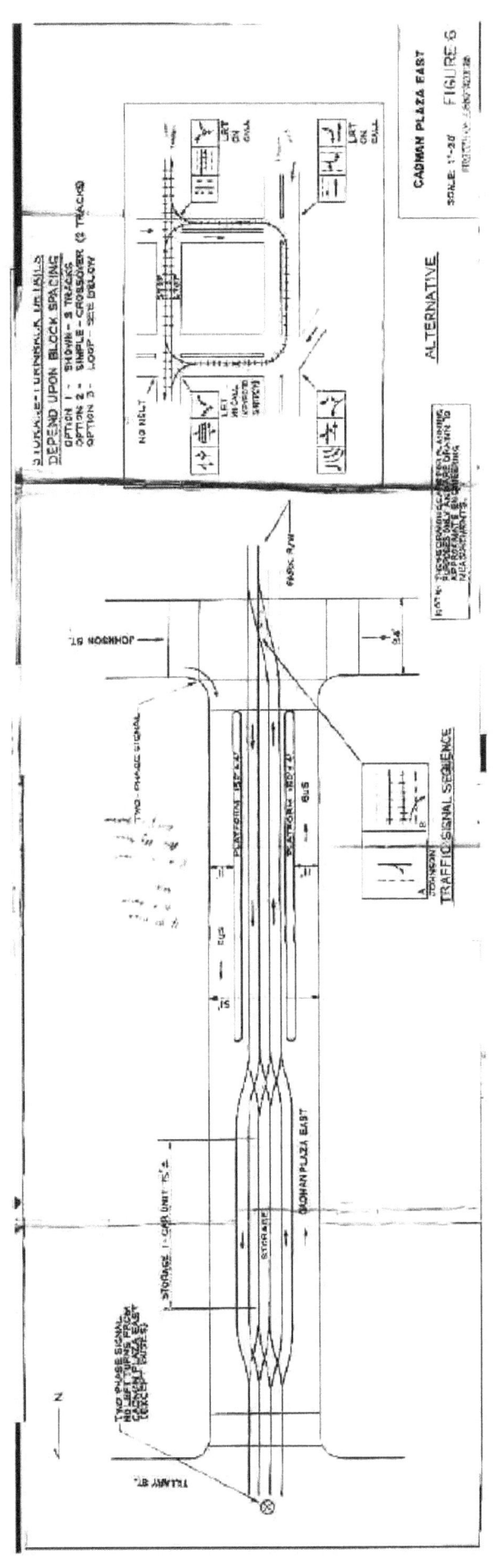

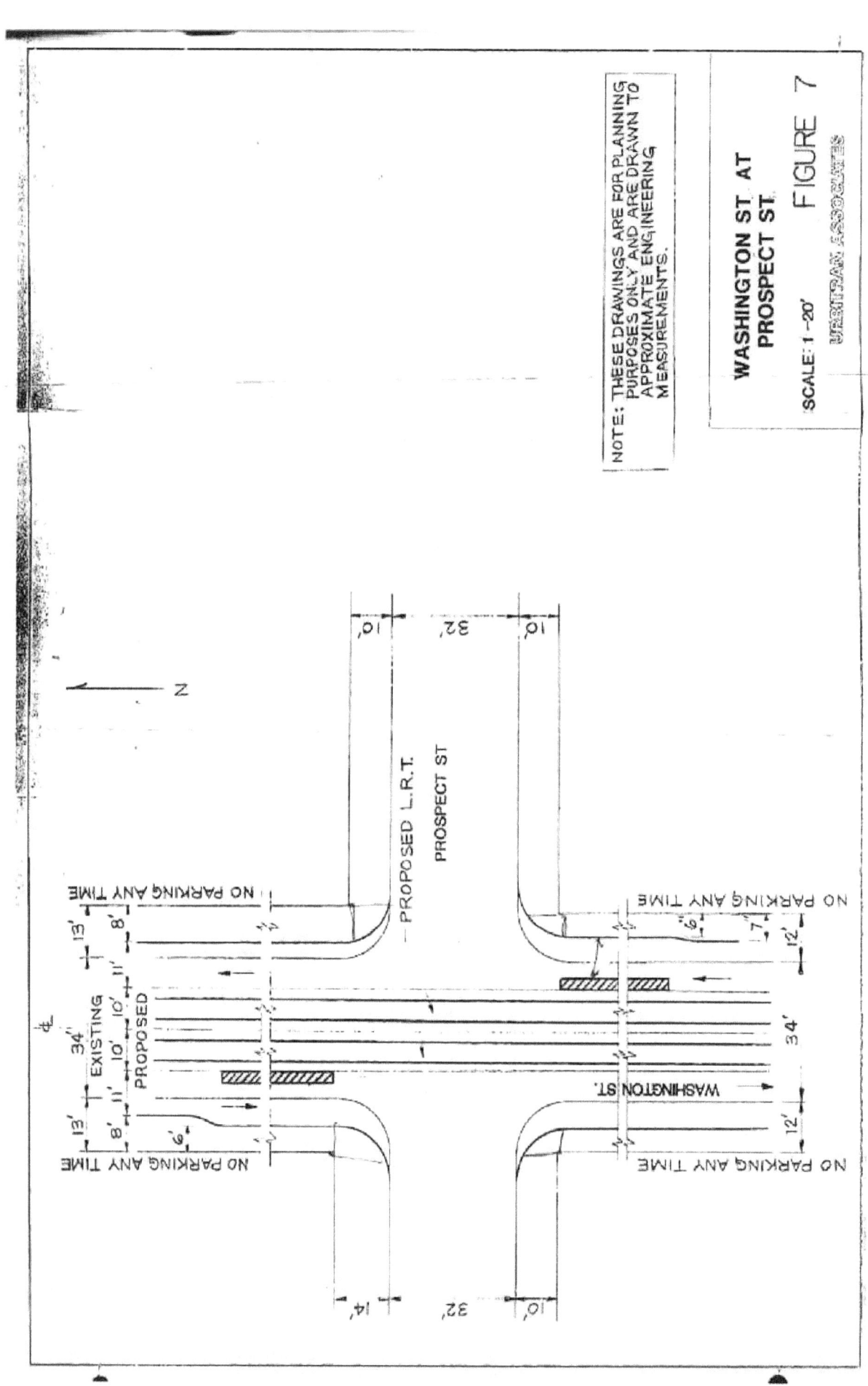

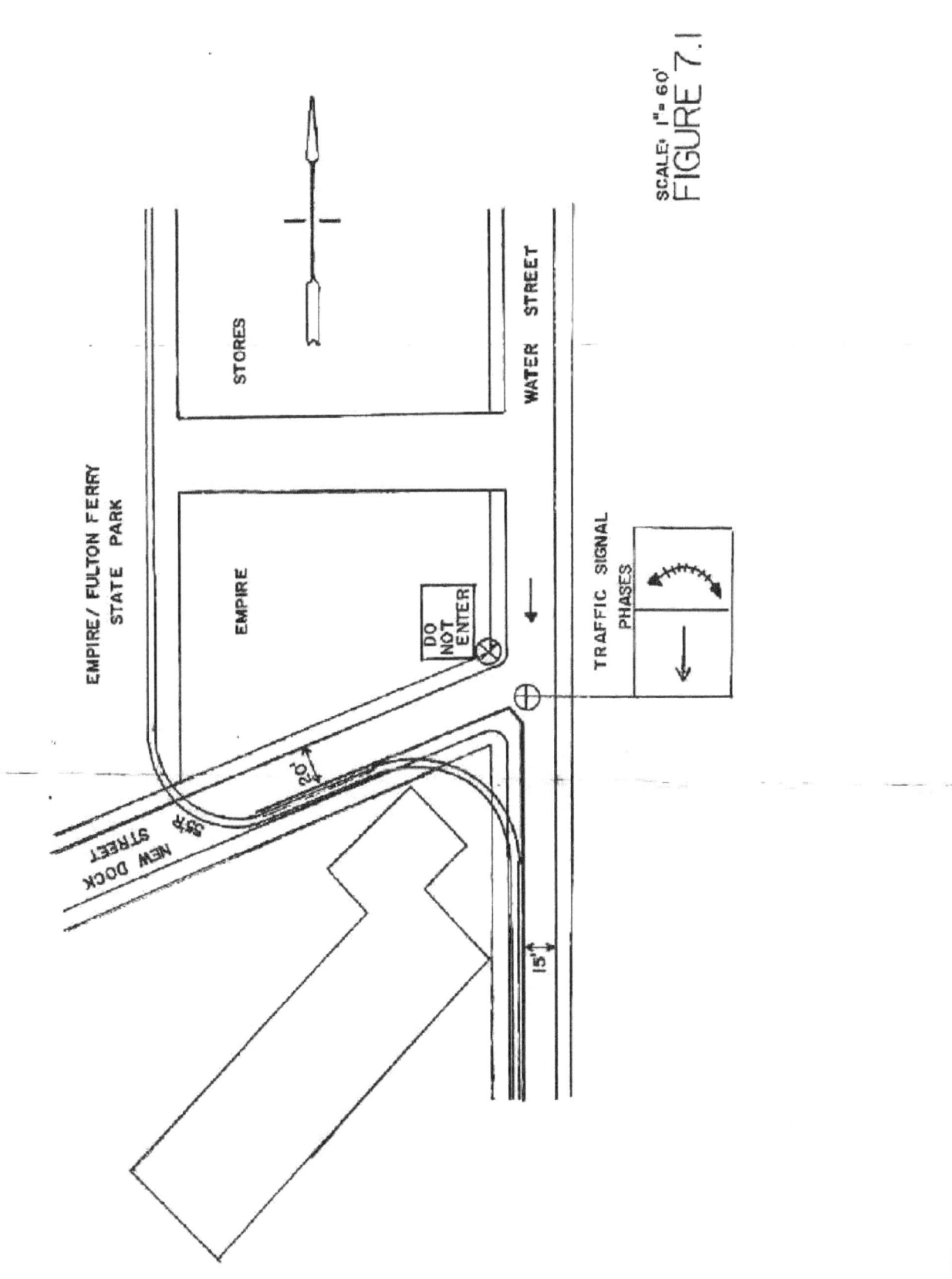

SCALE: 1"= 60'
FIGURE 7.1

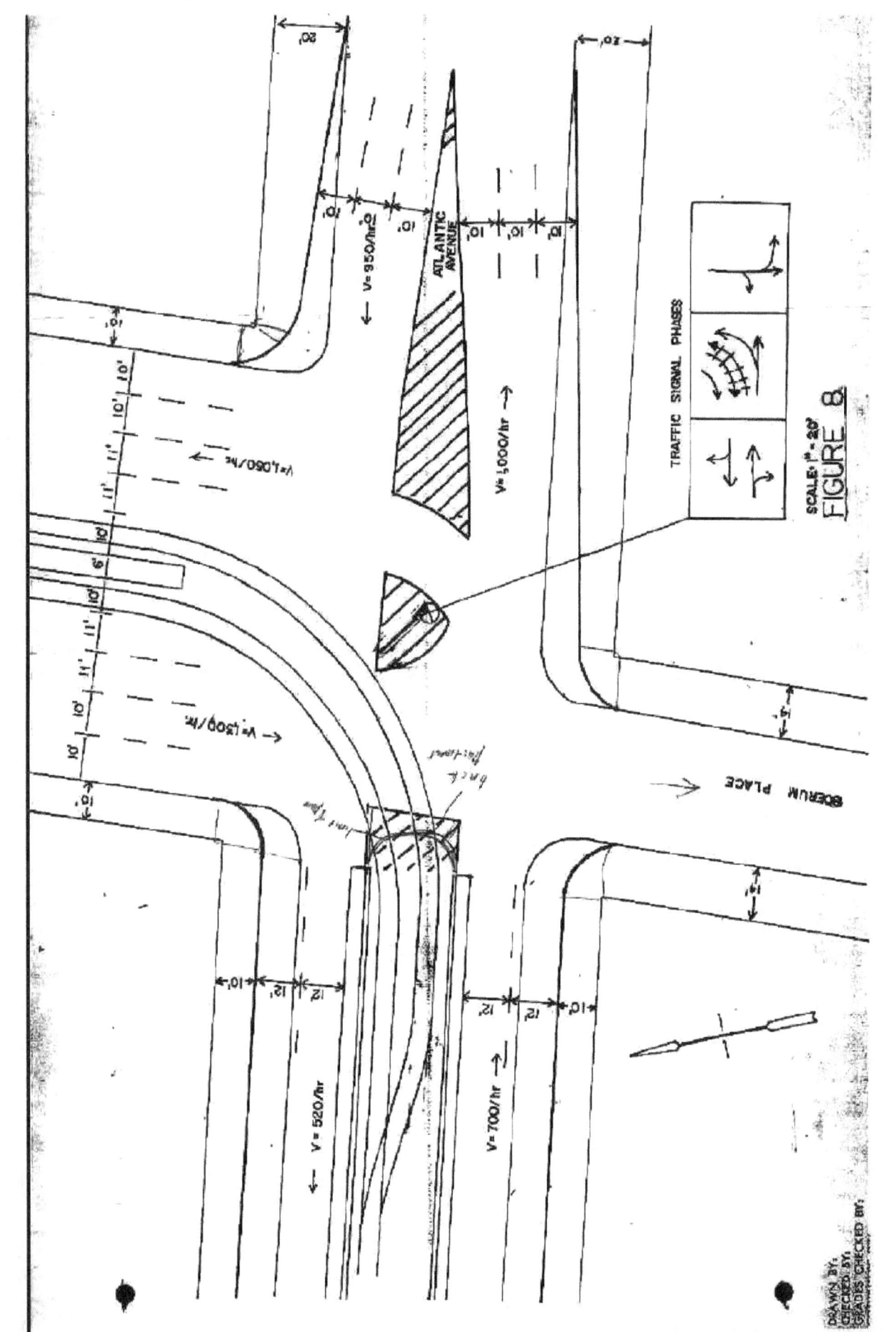

SCALE: 1" = 20'
FIGURE 8

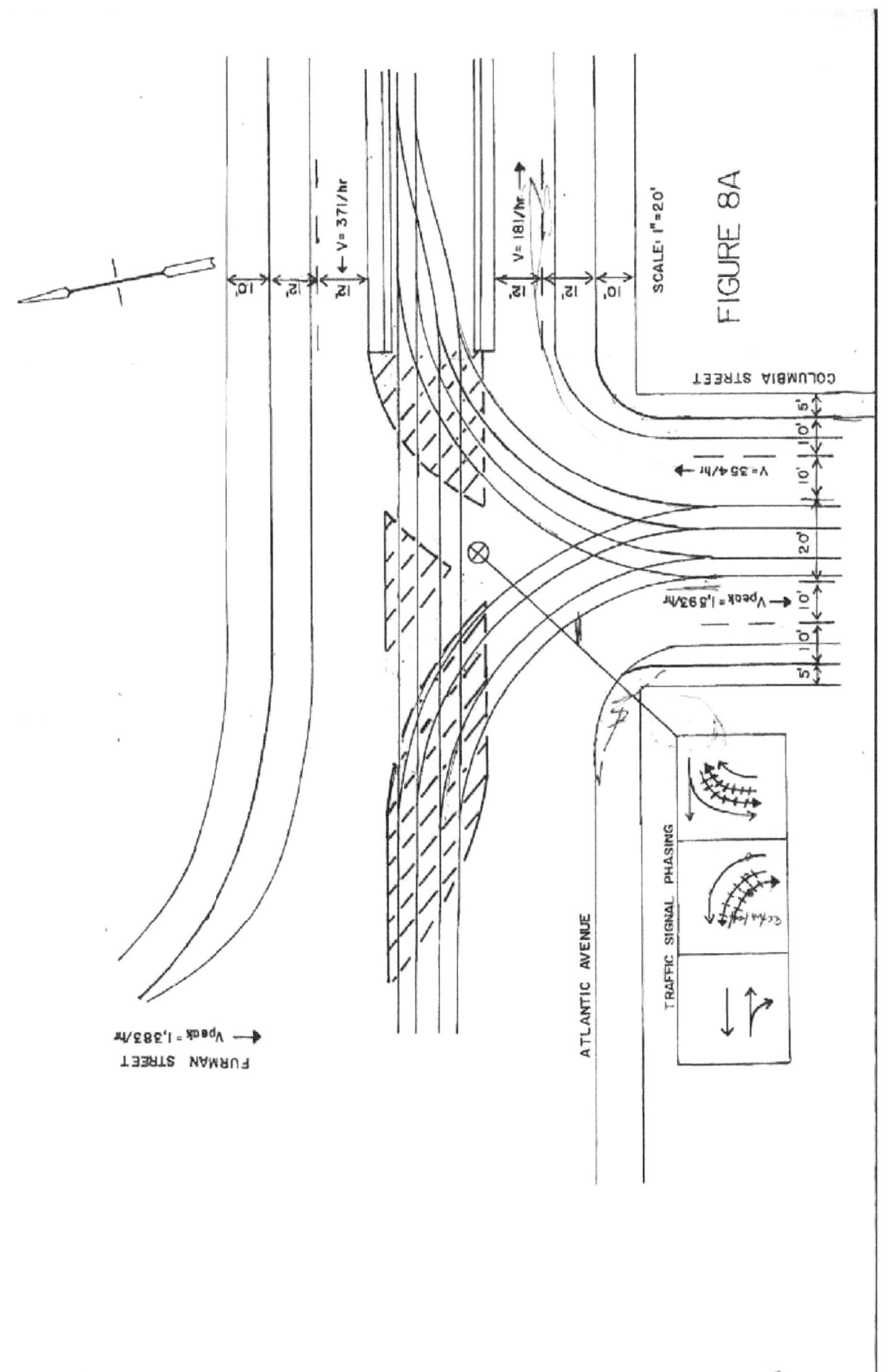

FIGURE 8A

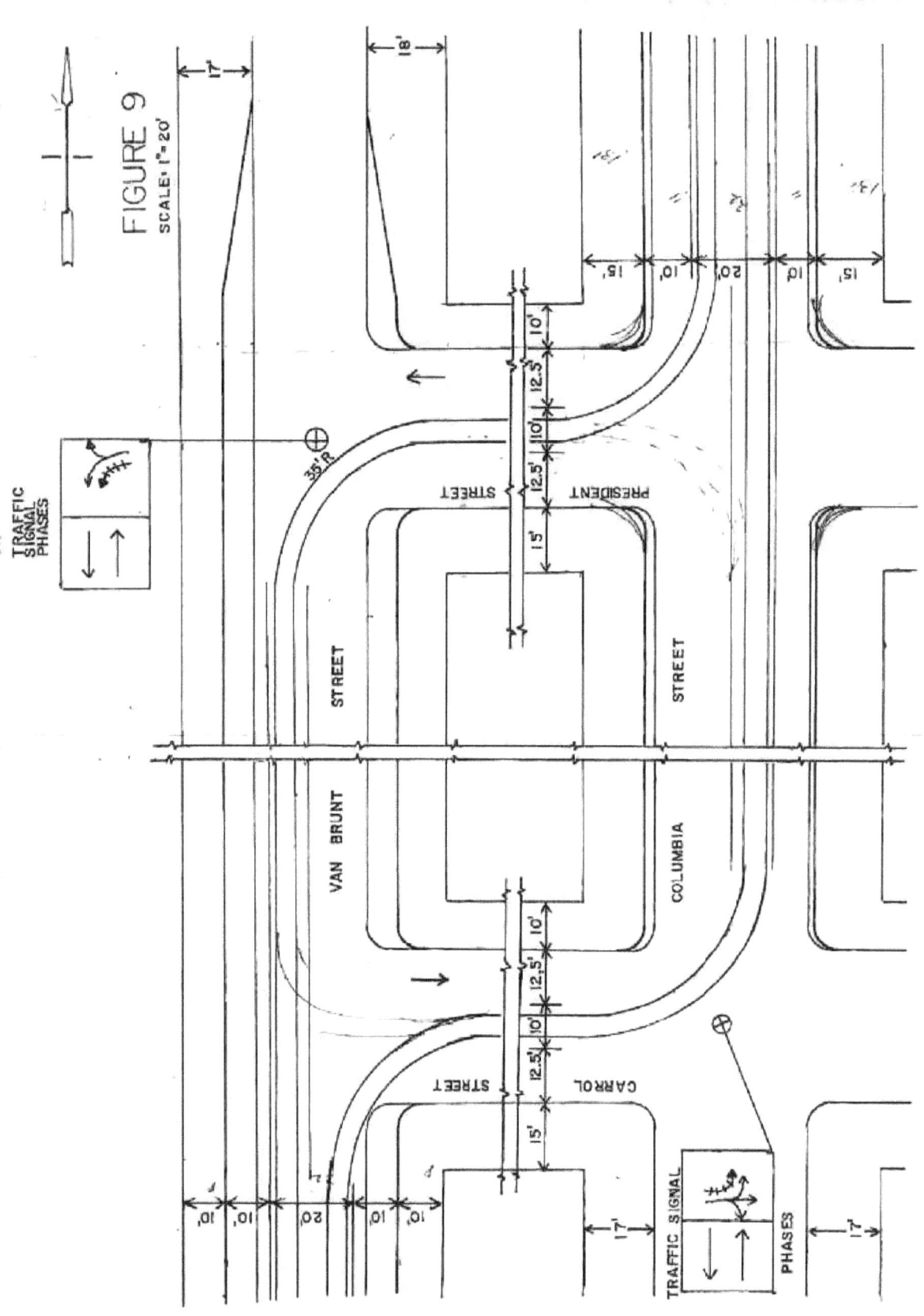

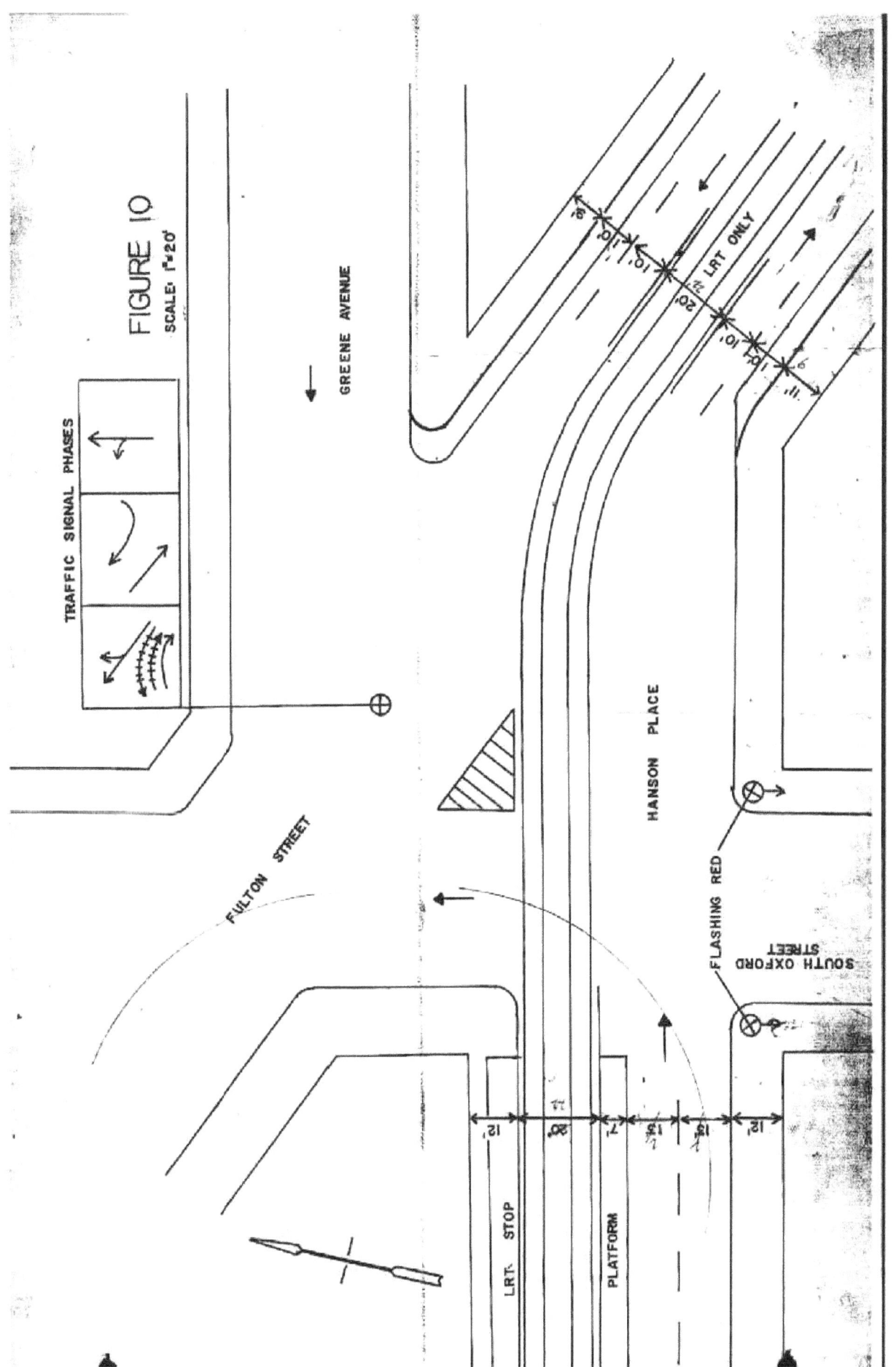

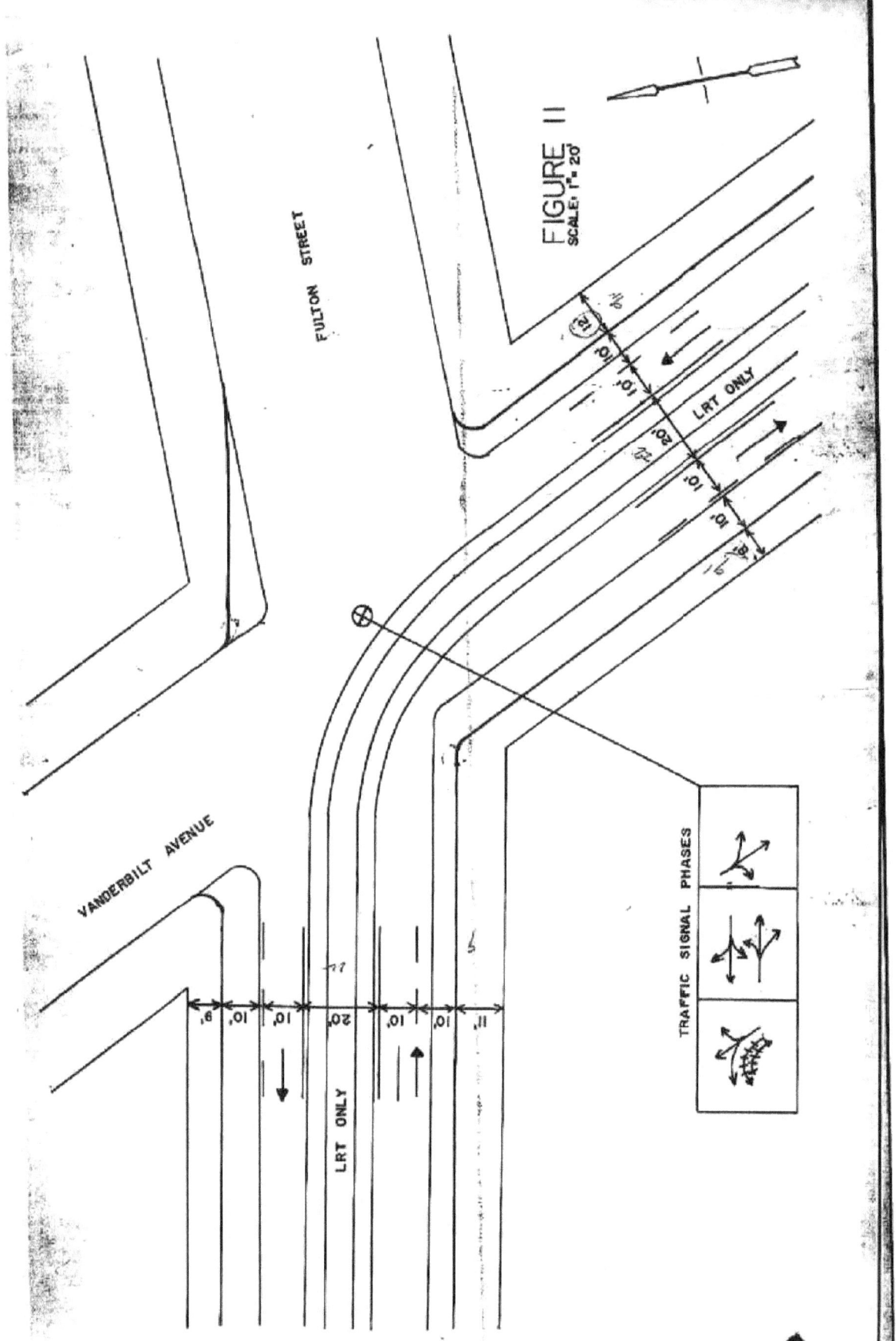

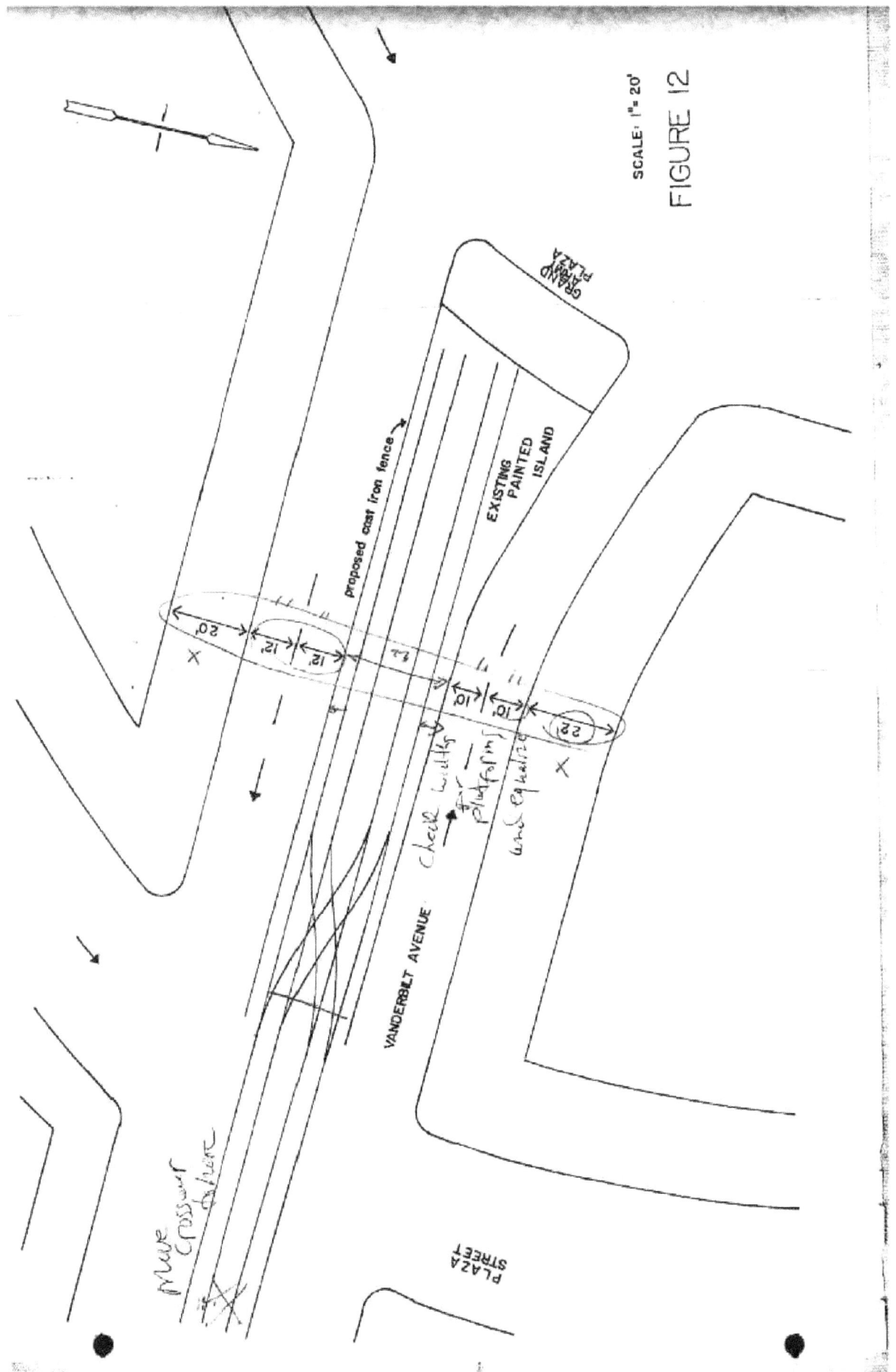

FIGURE 12

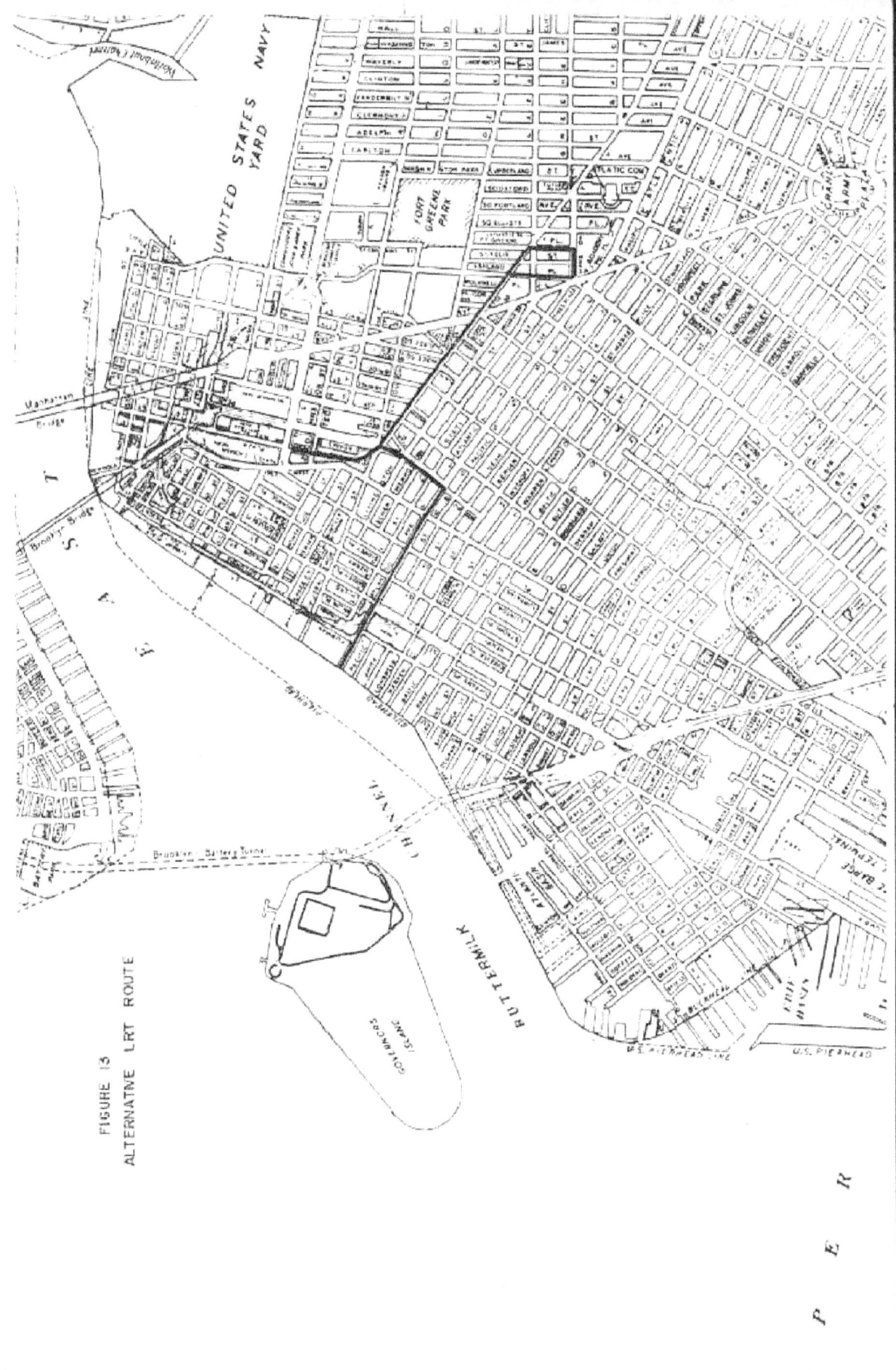

FIGURE 13
ALTERNATIVE LRT ROUTE

development projects which would be completed by this time would overload the subway system, even with planned increases in capacity. Overloading of the major arterial streets in downtown Brooklyn threaten to seriously impede bus service within and to the area. As an illustration, the following is quoted from the MetroTech EIS: "Given the scale of the proposed action (MetroTech), and the number of other major projects underway or proposed for downtown Brooklyn over the next five years, transportation systems in the area would be facing substantial inceases in overall demand and congestion". Further, "the area's parking supply would be tighter than today's conditions". Some of the traffic congestion impacts are identified as unmitigatable.

Phase one (1989) of MetroTech alone would increase subway use as follows :

    -AM increase : 4616 trips
    -PM increase : 4070 trips

    (see Appendix "F")

To quote the MetroTech EIS, "comparing existing and future planned capacity loads the additional passengers due to MetroTech Phase I would worsen scheduled capacity shortfalls on the IRT 4 and 5 lines outbound in the AM, the IRT 3 and 4 lines inbound in the AM and outbound in the PM, and outbound IND F line in the PM".

As for traffic congestion, and the attendant impediment to bus traffic, see Appendix "D" which lists key downtown Brooklyn intersections which will be overloaded by MetroTech alone.

Given this scenario, it is therefore reasonable to conclude that no additional potential mass transit users to downtown Brooklyn will be attracted away from their cars and onto mass transit.

Build Scenario, 1993:

Since the Light Rail routes in downtown Brooklyn would operate on lightly travelled streets through Fulton Mall and through a tunnel, it would be unaffected by area traffic congestion problems. Except for a short distance along Boerum Place from Atlantic Avenue to Fulton Mall where the LRV's would operate in dedicated lanes, the LRV routes are clearly separated from major flows to and from the Brooklyn and Manhattan Bridges.

As general automobile traffic problems in the City worsen, drivers will become more amenable to leaving their cars at home and use mass transit, provided of course, that there is a workable mass transit system available. For example, persons travelling to downtown Brooklyn by car from Long Island could leave their automobiles at the many Park and Ride facilities which the LIRR maintains, and take the LIRR to Atlantic Avenue and the Brooklyn Light Rail Link. Persons originating in New Jersey could take mass transit to Hoboken where ferry service would transport them to the Brooklyn Light Rail terminal at Atlantic Avenue. See Task 18 for a preliminary analysis. Residents of Red Hook, Brooklyn Heights, Cobble Hill, Fort Greene and parts of Park Slope who work in downtown Brooklyn or Manhattan, could leave their cars at home and use the Light Rail and Ferry service to and from their job and shopping sites.

Since the Light Rail routes parallel several bus routes, these bus routes could be cutback, thereby reducing the number of buses in downtown Brooklyn, with the attendant decrease in air pollution.

Bus Service Changes*

Some 15 local bus routes operate in downtown Brooklyn in the area bounded by Atlantic Avenue to the south and Flatbush Avenue to the east. These 15 routes will be impacted to a greater or lesser extent, by the Light Rail line.

The 15 bus routes have a wide range of origins outside the downtown area, and enter it via a number of roadways; from north to south and west, these include Nassau Street (Routes 57, 62, 69), Myrtle Avenue (Routes 54, 61), DeKalb Avenue (Route 38), Fulton Street (Routes 25, 26, 37), Flatbush Avenue (Routes 41, 45, 67), 5th Avenue (Route 63) 3rd Avenue (Route 37) and Smith Street (Route 75). Several of the routes do not penetrate into the Downtown area beyond its fringes; Route 63, which travels east-west on Atlantic Avenue; and Routes 57, 62, 69 all of which terminate at High Street Station.

*Note: parts of this section are based on the Final Report Transit Antic Study by Urbitran Associates in association with Seelye, Stevenson, Value & Knecht, 1985.

The routes most likely to be candidates for modification are those which parallel the Light Rail line and serve the heart of the downtown area, especially 8 routes: Routes 25, 26, 37, 38, 69, 61 and 52.

These 8 routes include the most heavily utilized route in Brooklyn, Route 41, with over 12 million riders in 1980; as well as the third most popular route, Route 38, with 5.2 million riders in 1980. The other four routes carried between 1.4 and 3.6 million riders each. The peak morning weekday headways on these routes ranges from 2 minutes for Route 41 to 9 minutes for Route 25. A route by route discussion follows -

Route 41 - Route 41 travels Livingston Street, Court Street and

TABLE 3 BUS OPERATIONS DATA

Route	Downtown Terminus and Streets	Weekends	Daily Hrs	AM Peak MF SA SU	Midday MF SA SU	PM Peak MF SA SU	Evening MF SA SU	Late NITE	1980 ridership (million)
19	Atlantic Terminal via DeKalb Ave	No	15	15 - -	- 30 -	- 30 -	- 35 -	-	2.4
25	Fulton Landing via Cadman Plaza West and Fulton Street	Yes	24	9 20 20	9 10 15	9 8 30	15 15 -	60	3.6
26	Adams/Johnson Sts via Fulton Street	Yes	24	7 15 15	9 7 10	8 8 10	13 18 18	45	2.9
37	York/Cold Sts. via Tillary/Adams/ Fulton/3rd Avenue	Yes	24	7 20 25	8 11 18	7 11 18	20 20 20	45	5.2
38	Tillary/Cadman Plaza West via Fulton	Yes	24	3 7 14	5 10	4 5 10	11 12 15	38	12.0
41	Fulton Landing via Livingston/ Cadman Plaza West	Yes	24	2 10 15	4 4 7	2 4 7	10 5 13	36	—
45	Court/Livingston via Livingston	Yes	24	5 11 15	7 5 15	5 5 15	10 10 15	60	3.4
52	Cadman Plaza West/Tillary via Fulton	Yes	24	4 6 15	5 6 12	5 6 12	12 10 15	64	—
54	Myrtle/Jay via Myrtle	Yes	24	4 10 20	6 6 7	4 6 7	6 12 12	30	4.4
57	High Street Subway Station via Sands/Nassau	Yes	24	5 15 20	10 12 12	6 12 12	15 20 20	60	1.9
61	Through route via Myrtle, Jay, Livingston and Atlantic.	Yes	24	7 12 20	7 8 18	7 8 18	12 18 18	60	—
62	High Street Subway Station via Sands/Nassau	Yes	24	8 11 15	8 10 15	8 10 15	15 20 15	50	1.9
63	Columbia St/Atlantic Avenue via Atlantic and 5th	Yes	24	6 12 15	7 6 6	6 6 5	6 15 12	45	6.7
67	Nassau/Jay via Jay/Livingston/ Flatbush	Yes	24	7 10 30	10 10 25	6 10 25	15 20 25	60	2.1
69	High Street Subway Station via Sands/Nassau	Yes	24	7 15 25	12 15 25	8 15 25	21 30 25	45	1.4
75	Nassau/Jay via Jay Street	Yes	24	7 15 30	9 11 25	7 11 25	22 20 25	56	2.1

Source: 1981 Brooklyn Bus Map, NYCTA

Cadman Plaza West. Most trips go to Fulton Ferry, but some are turned back at Johnson Street. This route could be cutback at Atlantic Terminal, as it duplicates the service of the Light Rail line.

Route 37 - Route 37 travels through downtown via 3rd Avenue, Flatbush Avenue, Fulton Street, Adams Street, and Tillary Street to the Brooklyn Navy Yard area. It would share Fulton Street with the Light Rail line as well as other bus routes, but unlike the other routes it extends in both directions well beyond the center of the downtown area. As such, this route cannot be terminated or cutback and replaced by the Light Rail. On the other hand, it could be moved off of Fulton Street onto Livingston Street, a recommendation made previously in the Downtown Brooklyn Traffic and Transit Study, to reduce congestion on Fulton Street and speed up transit operations.

Routes 25, 26, 52 - These three routes all use Fulton Street to enter the downtown area. Route 26 terminates in a loop on Adams Street, Tech Place, and Jay Street. Route 52 terminates in a loop around Borough Hall. Route 25 extends the full length of Fulton Street, and then proceeds north on Court Street and Cadman Plaza West to Fulton Landing. In essence, all three routes duplicate the service of the Light Rail line in the downtown area, except where Route 25 travels on Cadman Plaza West instead of Cadman Plaza East.

Operationally, these three routes would be terminated at Atlantic Terminal.

Route 38 - Route 38 operates via DeKalb Avenue inbound and Fulton Street/Lafayette Street outbound and ends in a loop on Court Street, Tillary Street, and Adams Street. This route would also be diverted to a terminal at Atlantic Terminal, by using Ashland Place south, with transfers required, but this diversion would add significantly to the travel time of bus riders, especially in the inbound direction.

Route 61 - Route 61 travels from Long Island City to Red Hook. In downtown Brooklyn it uses Myrtle Avenue to enter the Downtown area, and then operates along Jay Street, then along Atlantic Avenue, and then uses Columbia Street and Van Brunt Street to enter Red Hook. This route could be cutback at the foot of Atlantic Avenue as it duplicates Light Rail service into Red Hook.

Route 69 - Route 69 connects Grand Army Plaza with the Sands Street area. It uses Vanderbilt Avenue and Flushing Avenue. This route could be cutback at Fulton Street, or possibly eliminated, as its termination points are identical to one of the Light Rail routes, and part of the existing bus route is duplicated by other bus routes. This will require further study in the EIS.

Impacts and Mitigation:

The reduction of bus service will help reduce downtown Brooklyn air pollution levels.

Detailed planning will be needed to coordinate Bus and Light Rail schedules. Further, detailed design will be required for the proposed bus termination points at Atlantic Terminal.

Long Island Rail Road:

The LIRR operated 27 trains per peak A.M. hour, with a similar number of trains in the P.M. peak hours. The LIRR is not at capacity in its Brooklyn terminal. The LIRR can add more trains into their schedule, as per demand. The Light Rail project would provide this demand.

Ferries to downtown Brooklyn:

A ferry operation was started on a trial basis last year. It operated between Pier 17 and Old Fulton Street. It was not well used and disappeared after a few months. The ferry landing at Old Fulton Street was not readily accessable to residential and commercial centers in the area. The future of ferry service to downtown Brooklyn depends on having a mode of transportation that will move ferry passengers to and from their inland destinations in a speedy and comfortable fashion. Light Rail will provide this mode in downtown Brooklyn.

TASK 21

Construction Impacts

The City is planning street rebuilding projects for some of the streets along the proposed Light Rail routes. In particular :

HWK 715 Vanderbilt Avenue   1996

HWK 639 Cadman Plaza East   1996

HWK 700BW Columbia Street   1995

HWK 700A Van Brunt Street   1995

HWK 973 Fulton street   1996

The installation of the Light Rail track and overhead wire could be easily integrated into the street rebuilding, without appreciably increasing normal street reconstruction time schedules. With respect to noise and the maintenance of traffic, these issues would be accomodated as part of the normal street rebuilding process. No additional noise above the levels normally associated with a street construction would be created by the installation of the Light Rail tracks and wire. It would be necessary to realign a certain number of manholes and transformer vaults. This could be accomplished as part of the street rebuilding process.

The installation of Light Rail track in streets which are not scheduled for rebuilding would be accomplished as per Attachment "E". Bus and delivery truck traffic along Fulton Mall would be limited to single lane one-direction traffic only, while the track in the opposite lane would be installed. Buses operating in the opposite direction could be temporarily re-routed along Livingston Street until work in the corresponding lane in Fulton Mall would be completed. Planks over parts of track under construction would be installed to accomodate trucks making deliveries. Light Rail track can be installed at a rate of 1,000 feet per 5-day week. Taking into account the decorative pavement blocks used in the Mall, and the fact that only one track could be installed at a time, as well as the linear measurements of the Mall, it is estimated that each track could be installed in a period of two months, if the work was performed in a competent and diligent manner, which would be a total of four months for track construction within Fulton Mall. A key factor in speedy completion is the contractor maintaining a workable level of manpower at the site.

With regard to the Tunnel ramps, the sidewalk would first be reduced in width to 10 feet. The former sidewalk area would then be converted to roadway. The area within the perimeter of the ramp would then be excavated, while at the same time two moving lanes of traffic would be maintained.

Impacts and Mitigation:

Since track installation is similar in nature to street reconstruction, the impacts and mitigation for track installation will be similar to that of street reconstruction. If the track were installed as part of the aforementioned street rebuilding contracts, the impacts would be mitigated as part of the street reconstruction.

Time Required for Construction:

If the rate of 1,000 feet of track per week were maintained, and there were no deccelerating factors, the total construction time would be projected at 71 work weeks. 71 weeks represents the rate of work of a single construction crew. If additional crews worked simultaneously, construction time would be divided by the number of construction crews. This is based on the assumption that an adequate level of manpower will be maintained by the contractor at the job site until completion.

TASK 22

Alternative Plan:

The alternative to the primary plan is to phase in the feeder routes to Grand Army Plaza and Red Hook over an extended period of time. In this scenario, the LRT route configuration is indicated in Figure 13.

The results of this alternative would be as follows:

-The LRT routes would be reduced in length by 18,783 feet.

-Accordingly, construction time would be reduced.

-The B-69 and B-71 buses would retain their present routes. Reductions in air pollution due to replacement by LRT's on these routes would not take place.

-Some changes in parking regulations would not be needed.

-LRT service to the following recreational facilities, as well as their inclusion on the LRT loop would not take place: Fort Green Park, Brooklyn-Queens Greenbelt, Brooklyn Botanic Garden, Brooklyn Museum, Prospect Park, Atlantic Basin, Red Hook Park, Warehouse of William Beard, Red Hook Waterfront.

-Residents of Fort Green, Red Hook and part of Park Slope would not have LRT service.

-The redevelopment of Red Hook would face transportation shortfalls. The area would continue to remain isolated.

APPENDIX A

SINGSTAD, HURKA & ASSOCIATES, P.C.
Consulting Engineers

Frank R Hurka, P.E.
Anthony S. Caserta, P.E.
Joseph A. Delmo, R.A.
Stonslaw Szalowski, P.E.
Simon Zebrocky, P.E.

November 7, 1985

Brooklyn Historic Railway Association
599 E. 7th Street
Brooklyn, NY 11218

Re: Atlantic Avenue Tunnel
Engineering Report

Gentlemen:

In accordance with your request, I have completed the engineering analysis of the structural integrity of "The Atlantic Avenue Tunnel". The results of my investigation and the calculations are incorporated in the attached report entitled "Engineering Report - Atlantic Avenue Tunnel". This report is supplementary to my interim report of said structure dated August 7, 1984. (See Appendix A).

I have concluded that the existing tunnel structure is a safe and sound structure and that its structural integrity has not been compromised with age. The masonry and stone tunnel can sustain its present overburden load, and the live loads on Atlantic Avenue.

In my opinion, the useful life of the tunnel is not in jeopardy and the facility can be used to house an underground museum or exhibit hall.

Kindly advise if there are any questions or if I can be of any further assistance.

Respectfully submitted,

Anthony S. Caserta, P.E.

Attachments

SINGSTAD, HURKA & ASSOCIATES, P.C.

ENGINEERING REPORT
ATLANTIC AVENUE TUNNEL

Prepared by:

A.S. Caserta
P.E. No. 34498
NY State

SINGSTAD, HURKA & ASSOCIATES, P.C.

I. HISTORY OF THE TUNNEL

Construction on the railroad horse-shoe shaped tunnel started in May 1844 and was completed seven months later in December 1844. The entire structure was built by several hundred laborers using only hand tools such as pick, shovels and pack mules. The physical dimensions of the tunnel are impressive even by modern standards. The horseshoe shaped tunnel is twenty one (21) feet wide, seventeen (17) feet six (6) inches high and one thousand nine hundred and fifteen (1915) feet long. The tunnel has a brick arch roof section varying in thickness from 20 inches at the crown to 4 feet at the spring line. The stone masonry walls supporting the arch vary from 4 feet (spring line) to 6 feet at the base. (see Fig. 1).

The structure was built by the "cut and cover" method. Simply stated, this construction technique involves excavating from the surface grade down to the tunnel invert, erecting the side walls and roof arch, and then backfilling and repaving.

The tunnel was ventilated by the use of 3 large ventilation shafts extending to the street above. These ducts are oval-shaped with a maximum width of approximately 6 feet and are spaced 325 feet on centers along the middle third of the tunnel.

It is estimated that the tunnel was sealed up permanently in 1861 by the erection of 2 stone masonry walls inside of each portal; reducing the tunnel length to 1500 feet

II. TEST CORE SAMPLES AND TESTING

To complete the interim report dated August 7, 1984, the Brooklyn Historic Railway Association arranged to take core samples of the brick, stone and soil of the tunnel structure and have them tested.

The locations of the core samples were established by Mr. Caserta and selected to provide representative samples of the tunnel structure. (See Fig. 2). The test cores served two purposes as follows:

- Verify the thickness of the tunnel arch and wall sections

- Obtain compressive strength (p.s.i.) values of brick and stone by standard laboratory testing procedures

- Determine grain size analysis, sieve analysis and characteristics of existing soil in the invert and behind the tunnel structures.

The core drilling operation was done by the Semcor Inc. and completed on June 11, 1985.

SINGSTAD, HURKA & ASSOCIATES, P.C.

The core samples were deliever to the New York City Department of General Services Laboratory in July 1985. The tests outlined above were performed by the lab and completed on August 5, 1985 (See Appendix B).

III. STRUCTURAL ADEQUACY OF TUNNEL

As inspection was made of the tunnel by Mr. A. S. Caserta on July 19, 1984 and again on July 8, 1985. At both inspections, the masonry tunnel lining was observed to be:

- Free of any deterioration
- Showed no signs of distress
- Showed no signs of distortion
- Free of lossen brick or mortar

There was no presence of any water inflow at the invert or evidence of water infiltration through the masonary walls.

The laboratory report (NYC Dept. of General Services Laboratory) states that the compressive strength tests on the core samples yield an average value of 3151 p.s.i. (See Appendix B). Assuming an allowable compressive stress of 500 psi (0.15 f'c), it is apparent that the Atlantic Avenue Tunnel is satisfactory under the conservative overburden loading shown in the stress calculation.

The tunnel structural investigative analysis is based on several assumptions as follows:

1. The masonry tunnel lining has the lateral support of the surrounding sandy soil.

2. The tunnel lining thickness as determined by the cores is consistent for the entire tunnel length.

3. Since masonry cannot take tensile stress, the thrust (T) load is distributed over a smaller area.

The design overburden load on the lining is taken as the average depth of soil cover over the tunnel. Additionally, a live load of 500 p.s.f. for street traffic is superimposed in the overburden. For a design load of 10 feet of earth and a 500 p.s.f. live load, the calculations indicate a positive moment of 14,210 ft.lbs with a compressive thrust of 21,000 lbs per foot of tunnel at the spring line. Therefore the compressive stress in the masonry lining is approximately 390 p.s.i.

SINGSTAD, HURKA & ASSOCIATES, P.C.

The vertical side stone walls of the tunnel are bearing on non-cohesive soil and exerts a bearing load of 7300 p.s.f. at its foundation. The soil samples taken in the invert were tested and exhibit the characteristics of a granular soil without any clay content. According to the text "Foundation of Structures - Dunham, 3rd Edition", a conservative allowable bearing value for this material is 6000 to 8000 p.s.f.

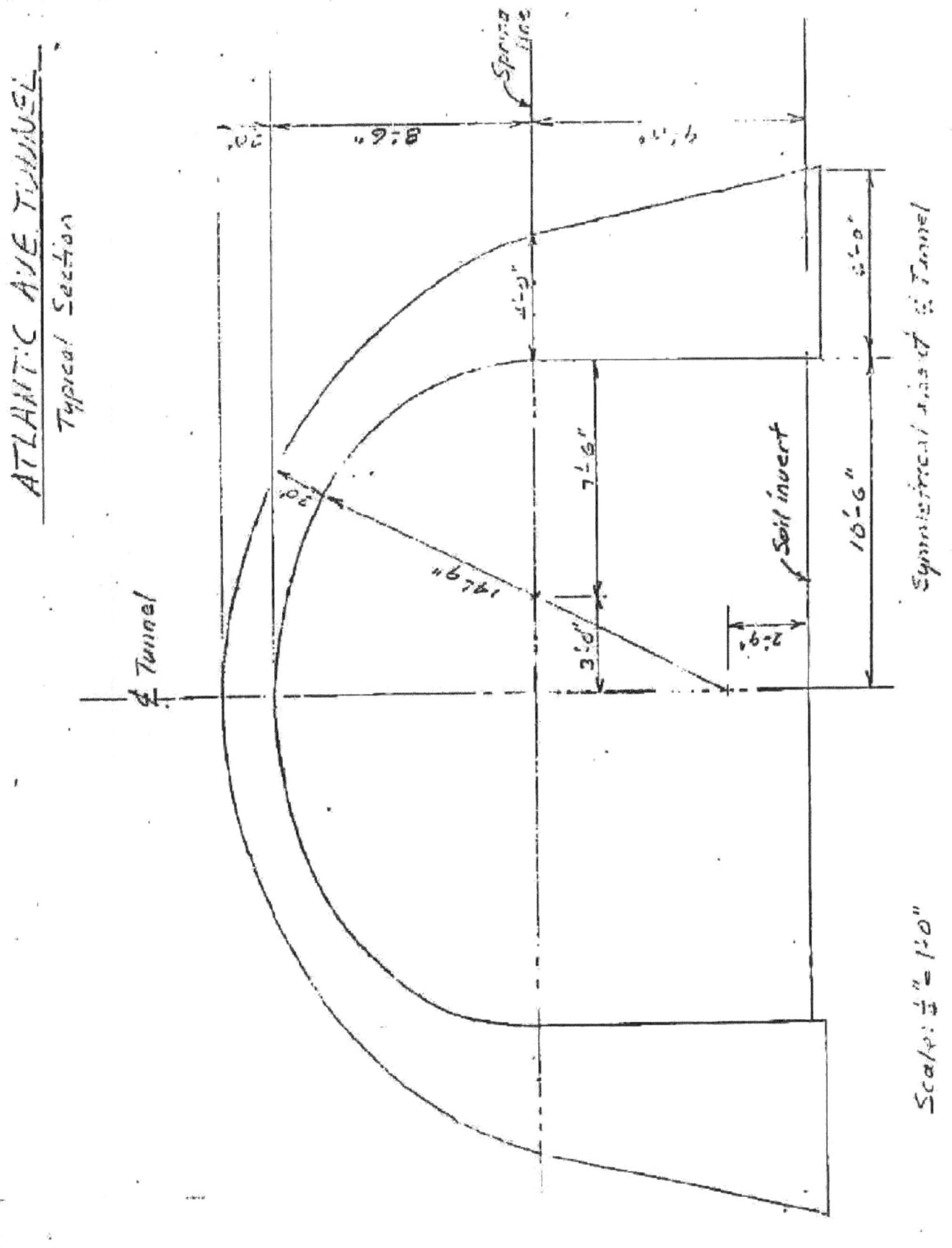

FIG. 1

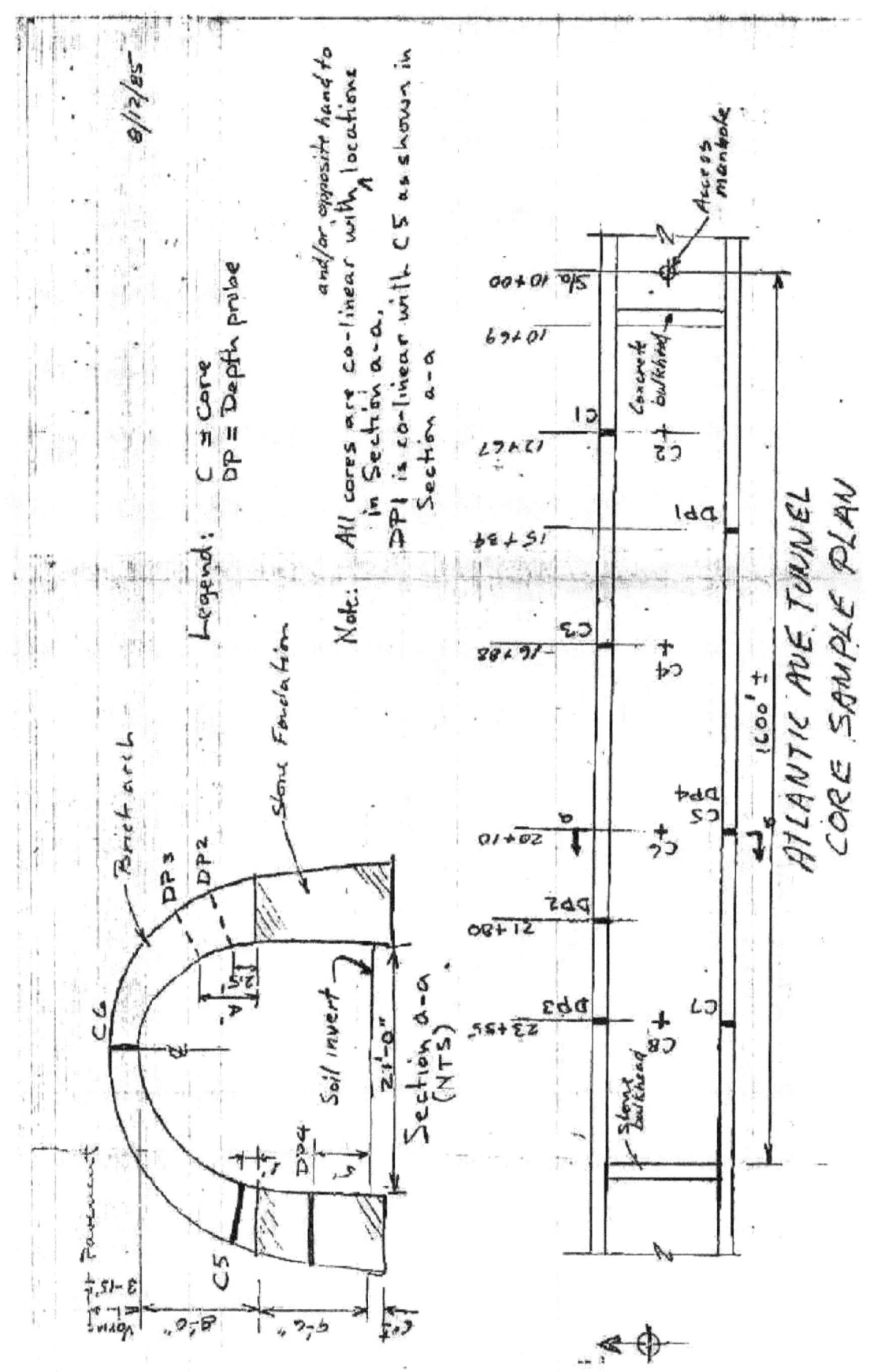

FIG 2

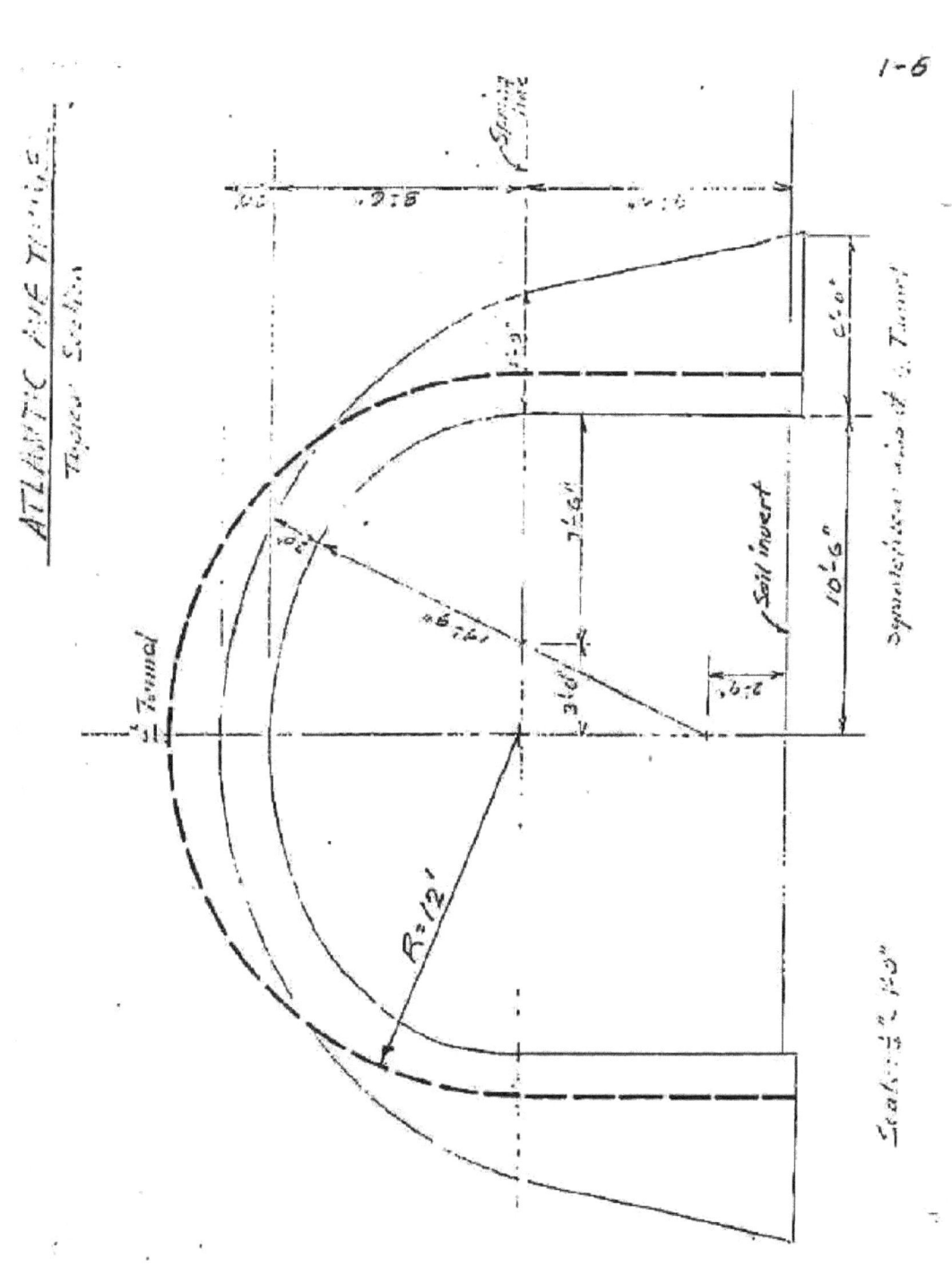

SINGSTAD, KEHART, NOVEMBER, & HURKA
CONSULTING ENGINEERS

PROJECT: Atlantic Ave Tunnel

$$M_\phi = -H(h + r\sin\phi) - Pr\sin\phi + \frac{wr^2}{2}\cos^2\phi$$

$H = -0.07123\,wr$
$P = 0.56761\,wr$

	0° S.L.	15°	30°	45°	60°	75°	90° Crown
$\sin\phi$	0	0.25882	0.50	.70711	.86603	.96593	1.0
$\sin^2\phi$	0	0.06699	0.25	.500	.75000	.93302	1.0
$r\sin\phi$	0	3.10584	6.00	8.48528	10.39236	11.59116	12.0
$r\sin^2\phi$	0	0.80388	3.00	6.00	9.00000	11.19624	12.0
$h + r\sin\phi$	9.5	12.10584	15.50	17.98528	19.89236	21.09116	21.5
$-H(h+r\sin\phi)$	+0.67669 wr	+0.8623 wr	+1.1040 wr	+1.28109 wr	+1.41693 wr	+1.50232 wr	+1.53145 wr
$-Pr\sin\phi$	−0	−1.76315	−3.40614	−4.81701	−5.89964	−6.58019	−6.81748
$+\frac{wr^2}{2}\cos^2\phi$ (noted as $+.5r\sin^2\phi$?)	+0	+0.40194	+1.5000	+3.00	+4.5000	+5.59814	+6.00000
$M_\phi =$	+0.67669 wr +8.1205 w	−0.48991 wr −5.98692 w	−0.80207 wr −9.6284 w	−0.53592 wr −6.43104 w	+0.01729 wr +0.20748 w	+0.52025 wr +6.2430 w	+0.71917 wr +8.6308 w

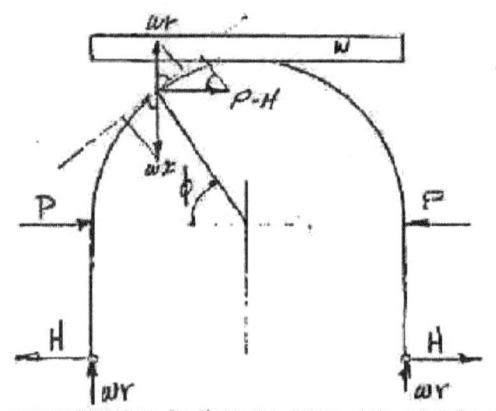

$$T = (wr - wx)\cos\phi + (P - H)\sin\phi$$
$$T = wr\cos^2\phi + (P - H)\sin\phi$$
$$(P - H) = (0.56769 - 0.07123)wr$$
$$= 0.49646\,wr$$

	0° S.L.	15°	30°	45°	60°	75°	90° Crown
$\sin\phi$	0	.25882	0.50	.70711	.86603	.96593	1.0
$\cos^2\phi$	1.0	0.93301	.75	.50	.25	.06699	0
$wr\cos^2\phi$	1.0 wr	.93301 wr	.75 wr	.50 wr	.25 wr	.06699 wr	0
$.49646\,wr\sin\phi$	0	.12849 wr	.24823 wr	.35105 wr	.42995 wr	.47955 wr	.49646 wr
THRUST =	1.0 wr / 12 w	1.06150 wr / 12.738 w	0.99823 wr / 11.97876 w	.85105 wr / 10.21260 w	.67995 wr / 8.15940 w	.54654 wr / 6.55848 w	.49646 wr / 5.9575 w

Singstad, Kehart, November, & Hurka
Consulting Engineers

Project: Atlantic Ave Tunnel
Sheet number: 3 of 6
Date: 10-21-92

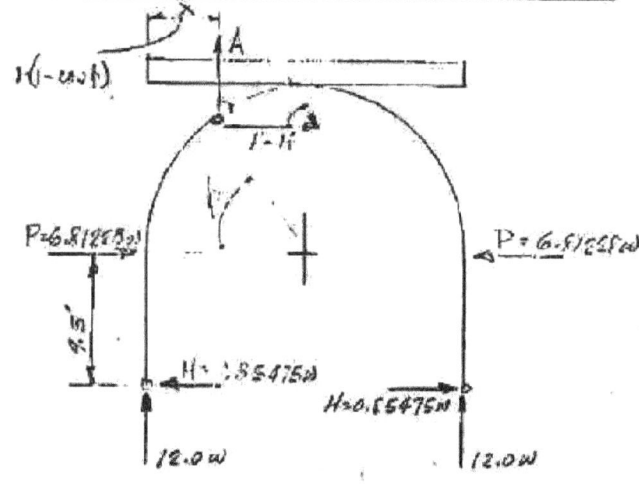

$1 - H = 0.4964...$ or
$= 5.95752\, w$

$A = 12w - wr(1-\cos\phi)$

$V = (P-H)\cos\phi - A\sin\phi$

	0° S.L	15	30	45	60	75	90 Cr
$\cos\phi$	1.0	.96593	.86603	.70711	.50	.25882	0
$\sin\phi$	0	.25882	.50	.70711	.86603	.96593	1.0
$r(1-\cos\phi)$	0	.40884	1.60764	3.51468	6.0	8.89416	12.0
A	12w	11.59116w	10.39236w	8.48532w	6.0w	3.10584w	0
$(P-H)\cos\phi$	5.93752w	5.74555w	5.15939w	4.21262w	2.97876w	1.54193w	0
$A\sin\phi$	0	3.00002w	5.19618w	6.00005w	5.17615w	3.00002w	0
V_ϕ	5.93752w	2.75453w	-0.03679w	-1.78743w	-2.21742w	-1.45809w	0

Uniform Vertical Load
@ 4' × 125 #/ft ; $\omega = 500$ #/ft
@ 6' × 125 #/ft ; $\omega = 750$ #/ft
@ 10' × 125 #/ft ; $\omega = 1250$ #/ft

SINGSTAD, KEMART, NOVEMBER, & HUSKA
CONSULTING ENGINEERS

PROJECT: ATLANTIC AVE. TUNNEL

FILE NUMBER: ___
SHEET NUMBER: 4 of __
DATE: 12-27-55
COMPUTED BY: ___
CHECKED BY: ___

SPRING LINE

	MOMENT (FT. LBS)				THRUST (LBS)			
	4'	6'	10'	15'	4'	6'	10'	15'
Uniform load	4065	6090	10,150	15,275	6000	9000	15000	22,500

@ 30° FROM S.L.

	MOMENT (FT. LBS)				THRUST (LBS)			
	4'	6'	10'	15'	4'	6'	10'	15'
Uniform Load	4812	7219	12031	18047	5989	8984	14,973	22,460

@ CROWN

	MOMENT (FT. LBS)				THRUST (LBS)			
	4'	6'	10'	15'	4'	6'	10'	15'
Uniform Load	4315	6473	10,788	16,181	2979	4468	7447	11,170

LIVE LOAD

$W_{L.L.} = 500$ psf is equivalent to 4' cover @ 125#/ft³

∴ For Investigation of Stresses — use average 10' + L.L.

Crown: $M_{@10'} = 10,788$ ft-lbs $T_{@10'} = 7447$ lbs
       $M_{L.L.} = \underline{4315}$ ft-lbs $T_{L.L.} = \underline{2979}$ lbs
              15,103 ft-lbs   10,426 lbs

Project: Atlantic Ave. Tunnel

@ 30° From C.L.

$M_D = -12,031$ $T_{D\,10'} = 14,973$
$M_{LL} = -4812$ $T_{LL} = 5981$
$\overline{-16,843 \text{ ft-lbs}}$ $\overline{20,762 \text{ lbs}}$

Spring Line

$M_{D\,10'} = +10,150$ $T_{D\,10'} = 15,000$
$M_{LL} = +4060$ $T_{LL} = 6,000$
$\overline{14,210 \text{ ft-lbs}}$ $\overline{21,000 \text{ lbs}}$

STRESS CALCULATIONS

$$f_c = \frac{T}{A} \pm \frac{6M}{bd^2} \; ; \; f_{c\,max} = \frac{2T}{3bg} \text{ if tensile stress exists}$$

15' Cover w/LL	M ft-lbs	T lbs	b	d	d²	f_c (psi)	f_t (psi)	g	$f_{c\,max}$ (psi)
Sp. Line	14,210	21,000	12	20	400	27.5	213.2	3	383.7
30°	-16,843	20,962	12	20	400	87.3	252.6	3	388.2
Crown	15,103	10,426	12	20	400	43.4	226.5	5	193.1

SINGSTAD, KEHART, NOVEMBER, & HURKA
CONSULTING ENGINEERS

PROJECT: Henry Avenue Tunnel

BEARING VALUES

$35' \times 8'(avg) \times 125 \, psf = 41,250 \, lbs/ft$

$\frac{54'}{2} \times 2.83' \times 130 \, pcf = 12,875$

$(+25 \times 3'(avg) \times 130 \, psf = 11,700$

LIVE LOAD   500 × 33   = $17,000$

$87,995 \, lbs/ft$

Bearing Pressure = $\dfrac{87,995 \, lbs/ft}{12 \, ft} = 7300 \, lbs/ft^2$

SINGSTAD, HURKA & ASSOCIATES, P.C.
Consulting Engineers

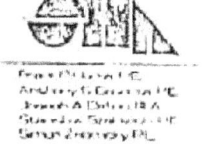

August 7, 1984

Brooklyn Historic Railway Association
599 East 7th Street
Brooklyn, New York 11218

Attention: Mr. Robert Diamond, President

RE: <u>The Atlantic Avenue Tunnel</u>

Gentlemen:

On Thursday, July 19, 1984, at your request, the undersigned and Mr. J. A. Defino (Chief Architect, SHA) participated with members of the Brooklyn Historic Railway Association (BHRA) in a preliminary survey to determine the existing condition of "The Atlantic Avenue Tunnel" which was constructed approximately one hundred and forty years ago.

The group entered the abandoned tunnel through an access manhole (at the corner of Atlantic Avenue and Court Street) and proceeded to conduct a visual and general survey of the tunnel; from the existing east bulkhead (west of Court Street) to the west bulkhead (east of Columbia Street), a distance of approximately 1500 feet.

The tunnel cross section is an horseshoe shaped structure approximately 21 feet wide at the base and 17 feet high at the centerline. It is comprised of granite block vertical walls to the springline and then brick masonry for the arched portion of the horseshoe.

An instrumental survey and detailed inspection of the tunnel were not possible due to the limited time available to inspect the site. However, sufficient information was obtained from visual observations to permit the following conclusions:

APPENDIX "A"

SINGSTAD, HURKA & ASSOCIATES, P.C.

Attention: Mr. Robert Diamond   August 7, 1984

1. For a tunnel structure which is approximately 140 years of age, the observed portion of the masonry tunnel (1500 feet long) is remarkably free of deterioration and appears to be structurally sound.

2. There is no presence of any inflow of ground water at the invert and the tunnel interior is relatively dry and shows no signs of any water infiltration.

3. There is evidence of moisture droplets on the masonry arch and the fluorescent light fixtures in certain portions of the tunnel. This is due to atmospheric condensation created by the tunnel "opening" in 1980. This can be corrected by installing mechanical ventilation fans to permit fresh air circulation throughout the tunnel.

It should be noted that while the apparent condition of the masonry tunnel and the quality of the workmanship indicates extraordinary technical competence, the state of the art of tunnel design was in its infancy when this tunnel was constructed. An in depth design analysis, using field measured data, would be needed for a definitive determination of any rehabilitation need.

In conclusion, insofar as a preliminary, visual survey can determine, "The Atlantic Avenue Tunnel" is presently structurally sound and quite safe for authorized personnel to enter thereon to conduct further detailed investigations.

Singstad, Hurka & Associates is quite pleased to be part of the restoration program of "The Atlantic Avenue Tunnel" and is ready to assist BHRA in whatever way it can to fulfill the goal of revitalizing Atlantic Avenue and its immediate community.

Very truly yours,

SINGSTAD, HURKA & ASSOCIATES, P.C.

Anthony S. Caserta
Executive Vice President

ASC:fk
cc: Mr. Sebastian Scialabba, P.E.

CITY OF NEW YORK
DEPARTMENT OF GENERAL SERVICES
DIVISION OF MUNICIPAL SUPPLIES
Laboratory
INTRA-DEPARTMENT
MEMORANDUM

TO: Robert Diamond, President
Brooklyn Historic Railroad Association   DATE: Aug 07 ' 85

FROM: A.D. Pacifico, DGS Laboratory Director
SUBJECT: Test Results   9 2 5 - 5 4 0 6

Attached is our laboratory report on samples of cores, brick, and soil taken from the Atlantic Avenue Tunnel.

Professionally, we were glad to have the opportunity to test materials which were used in a construction project dating back to the mid 19th Century.

We hope the data supplied is sufficient for your needs. If additional information is required, please contact me.

We have set aside the test samples and await your instructions, as to their disposition.

Attachment: Lab Report B6 - 0886
ADP: ejw

A QUALITATIVE REVIEW OF CORES RECEIVED FROM THE ATLANTIC AVENUE TUNNEL IN BROOKLYN

THE FOLLOWING IS A BRIEF DESCRIPTION OF CORES RECEIVED BY THIS LABORATORY FOR EVALUATION. THE CORES WERE TAKEN FROM VARIOUS LOCATIONS IN THE TUNNEL AT ATLANTIC AVENUE BETWEEN HICKS AND COURT STREETS, BROOKLYN.

/CORE #1/
Core #1 is composed of a brick section and a stone section. In the brick section, there are 5 layers of brick with mortar joints, which measure approximately 20 inches. The mortar bond still exixts between each individual brick. Two composite brick samples were taken from this core for test. The results are attached. The stone section has pieces of random lengths. There is one piece, 15 inches, and another 12 inches. The others are fragments of varying sizes from 3 inches to small pebbles. This core was taken from the side of the Tunnel.

/CORE #2/
Core #2 has a brick section only. There are 5 layers of brick with mortar joints. The total length is approximately 20.5 inches. The individual bricks are still bonded together with mortar. One composite brick sample was tested from this section, and one individual brick sample. The test results are attahced. This core was taken from the ceiling of the Tunnel.

/CORE #3/
Core #3 is of brick section and stone section. There are 5 layers of brick with mortar joints and these measure approximately 20 inches. The individual bricks are bonded by mortar joints. In the stone section, there is one piece, 13.5 inches, another 12 inches, and fragments and pebbles of varying sizes. No test sample was taken from this core. The core was obtained from the side of the Tunnel.

/CORE #4/
Core #4 is of brick section only. There are 5 layers of brick with mortar joints measuring approximately 20 inches. The mortar bond still exists between the individual bricks. No test sample was taken from this core. The core was obtained from the ceiling of the Tunnel.

QUALITATIVE REVIEW OF CORES RECEIVED FROM ATLANTIC AVE. TUNNEL - BKN

RE: CORES .... c o n t i n u e d ...

/CORE #5/
Core #5 is of brick section and stone section. Brick section measures approximately 20 inches. There are 5 layers of brick with mortar joints. The bond between individual bricks still exists. The stone section has various sizes of stone held together with mortar and measures approximately 22 inches. One composite brick sample and one individual brick were tested from this core. The results are attached. This core was taken from the side of the Tunnel.

/CORE #6/
Core #6 is of brick section only. There are 5 layers of brick bonded by mortar. Individual bricks are also bonded by mortar. Two composite brick samples were tested from this core. This core was taken from the ceiling of the Tunnel.

/CORE #7/
Core #7 is comprised of brick section and stone section. The brick section has 5 layers of brick with mortar joints and measures approximately 20 inches. No test sample was taken from this core. The stone section has an overall length of approximately 23 inches. One piece is 17 inches, and the other pieces are of varying sizes held together by mortar. This core was taken from side of Tunnel.

/CORE #8/
Core #8 is of brick section only. There are 5 layers of brick with mortar joints. Overall measurement is approximately 20 inches. Each individual brick is bonded with mortar. No test sample was taken from this core. The core was obtained from ceiling of Tunnel.

/CORE #DP4/
Core #DP4 is of varying sizes of stone, held together by mortar. The overall length is approximately 62 inches. The stone appears to be Mica Schist, based upon observation of its structure and the amount of mica flakes contained.

Two samples were taken for degradation test by Los Angeles Machine, as per A.S.T.M. C 535.

The results of the tests are attached.

CITY OF NEW YORK
DEPARTMENT OF GENERAL SERVICES
LABORATORY - 480 Canal St., NYC

Sheet #3 of 13 sheets

Date 7-16-85

CITY OF NEW YORK
DEPARTMENT OF GENERAL SERVICES
Bureau of Laboratories

Insp. # _____ Lab. # _____
Dept. _____ Sect. _____
Material 4" Ø BRICK CORE   Spec. # _____
Brand _____ Contract # _____
Contract ATLANTIC AVE. TUNNEL
Point/Date of Sampling
Remarks CORE #1
Sampled by: Robert Diamond
                    (Signature)

Lab # B6-0886

Insp. # _____

Contract # _____

79 - Date Rec'd by Lab.

Specification   Date received in lab 6 - 85

TESTS	Found	Requirements
WATER CONTENT, 24 hrs. IN COLD WATER (%)	10.1	
WATER CONTENT, 5 hrs. BOILING, (%)	12.1	
COMPRESSIVE STRENGTH (PSI):		
SAMPLE # 1 COMPOSITE	3,382	
SAMPLE # 1A COMPOSITE	2,626	
SAMPLE #		
SAMPLE #		
SAMPLE #		

CITY OF NEW YORK
DEPARTMENT OF GENERAL SERVICES
LABORATORY - 480 Canal St., NYC

Sheet #4 of 13 Sheets

Date 7-16-85

CITY OF NEW YORK
DEPARTMENT OF GENERAL SERVICES
Bureau of Laboratories

Lab # B6-0886

Insp. # _____ Lab. # _____
Dept. _____ Sect. _____
Material 4" Ø Brick Core Spec. # _____   Insp. # _____
Brand _____ Contract # _____
Contract ATLANTIC AVE. TUNNEL   Contract # _____
Point/Date of Sampling _____
Remarks CORE # 2
Sampled by: ROBERT DIAMOND
(Signature)

79 - Date Rec'd by Lab.

Classification   Date received in lab 7-85

TESTS	Found	Requirements
WATER CONTENT, 24 hrs. IN COLD WATER (%)	10.2	
WATER CONTENT, 5 hrs. BOILING, (%)	11.3	
COMPRESSIVE STRENGTH (PSI):		
SAMPLE # 2 COMPOSITE	3,581	
SAMPLE # 2 HALF BRICK	3,076	
SAMPLE #		
SAMPLE #		
SAMPLE #		

CITY OF NEW YORK
DEPARTMENT OF GENERAL SERVICES
LABORATORY - 480 Canal St., NYC

Sheet #5 of 13 sheets

Date 7-16-85

CITY OF NEW YORK
DEPARTMENT OF GENERAL SERVICES
Bureau of Laboratories

Insp. # _____ Lab. # _____
Dept. _____ Sect. _____
Material 4" ⌀ Brick Core   Spec. # _____
Brand _____ Contract # _____
Contract ATLANTIC AVE. TUNNEL
Point/Date of Sampling _____
Remarks CORE #5
Sampled by: Robert Diamond
(Signature)

Lab # B6-0886

Insp. 0 _____

Contract # _____

79 - Date Rec'd by Lab.

Specification   Date received in lab 7 - 85

TESTS	Found	Requirements
WATER CONTENT, 24 hrs. IN COLD WATER (%)	11.3	
WATER CONTENT, 5 hrs. BOILING, (%)	12.5	
COMPRESSIVE STRENGTH (PSI):		
SAMPLE # 5 — COMPOSITE	3,242	
SAMPLE # 5A — HALF BRICK	3,451	
SAMPLE #		
SAMPLE #		
SAMPLE #		

CITY OF NEW YORK
DEPARTMENT OF GENERAL SERVICES
LABORATORY - 480 Canal St., NYC

Sheet #6 of 13 sheets

Date 7-16-85

CITY OF NEW YORK
DEPARTMENT OF GENERAL SERVICES
Bureau of Laboratories

Lab # BE-0886

Insp. # _____ Lab. # _____
Dept. _____ Sect. _____
Material 4" ⌀ Brick Core Spec. # _____   Insp. # _____
Brand _____ Contract # _____
Contract ATLANTIC AVE. TUNNEL   Contract # _____
Point/Date of Sampling _____
Remarks CORE #6
Sampled by: Robert Dinmond
                (Signature)

79 - Date Rec'd by Lab.

Specification   Date received in lab 7- 85

TESTS	Found	Requirements
WATER CONTENT, 24 hrs. IN COLD WATER (%)	10.3	
WATER CONTENT, 5 hrs. BOILING, (%)	11.9	
COMPRESSIVE STRENGTH (PSI):		
SAMPLE # 6 COMPOSITE	3,183	
SAMPLE # 6A COMPOSITE	2,667	
SAMPLE #		
SAMPLE #		
SAMPLE #		

DEPARTMENT OF GENERAL SERVICES
LABORATORY - 480 Canal St., NYC

Sheet # 1 of 12 sheets
Date 7-16-85

CITY OF NEW YORK
DEPARTMENT OF GENERAL SERVICES
Bureau of Laboratories

I.#_____ Lab.# D6-0886
Dept._____ Sect._____
Material STONE Spec.#_____
Brand_____ Contract #_____
Contract ATLANTIC AVE. TUNNEL
Point/Date of Sampling_____
Remarks LOS ANGELES MACHINE TEST
Sampled by: ROBERT DIAMOND
(Signature)

Lab. # D6-0886

Insp. #_____

Contract #_____

INFORMATION

79 - Date Rec'd by Lab.

Specification   Date received in lab. 7- 85

Tests	Found	Requirements
LOS ANGELES MACHINE TEST		
1 - Loss in weight for sample with maximum size of 3 inches	32.45 %	
2 - Loss in weight for sample the same size as core was received	7.58 %	

Note: The Los Angeles Machine Test was run on both the stone sample as received and the stone sample broken into approximately 3" pieces.
The higher result for the 3 inch pieces can be explained by the increased surface area and the softer interior of the sample.

(7)

CITY OF NEW YORK
DEPARTMENT OF GENERAL SERVICES
LABORATORY - 480 Canal St., NYC

Sheet #8 of 13 Sheets

Date 7/31/85

CITY OF NEW YORK
DEPARTMENT OF GENERAL SERVICES
Bureau of Laboratories

Insp. # _____ Lab. # _____
Dept. General Services Sect. _____
Material Soil
Brand _____ Spec. # _____
Contract _____ Contract # _____
Point/Date of Sampling Old subway tunnel at Atlantic
Remarks Avenue & Court Street
Sampled by: Joseph
(Signature)

Lab. # BE-0886
Sample 1
Insp. # _____
Contract # _____

INFORMATION

SAMPLE 1 WAS TAKEN FROM CORE LOCATION No. 8

79 - Date Rec'd by Lab. JUL 23 1985

Eleanor C. Eastman

Specification   Date received in lab.

Tests	Found	Requirements
Material	Soil	
Appearance	MEDIUM BROWN IN COLOR CONTAINS STONES OF VARIOUS SIZES	
Natural Water Content, %	3.5	
Sieve Analysis, By weight passing, % Sieve Size:		
1 inch	53.3	
3/8 inch	47.0	
No. 4	42.5	
No. 8	38.0	
No. 40	19.4	
No. 200	4.6	
Plastic Limits, %	NONPLASTIC	
	(8)	Paul Kruger

Sheet #9 of 13 sheets

DEPARTMENT OF GENERAL SERVICES
LABORATORY - 480 Canal St., NYC

Date 7/31/85

CITY OF NEW YORK
DEPARTMENT OF GENERAL SERVICES
Bureau of Laboratories

Lab. # D6-0886
Sample 2
Insp. #

Insp. # _____ Lab. # _____
Dept. General Services Sect. _____
Material Soil   Spec. # _____
Brand _____ Contract # _____
Contract _____
Point/Date of Sampling Old subway tunnel at Atlantic
Remarks Avenue & Court Street
Sampled by: Joseph
                  (Signature)

Contract # _____

INFORMATION

SAMPLE 2 WAS TAKEN FROM CORE LOCATION No. 7
79 - Date Rec'd by Lab. JUL 2 3 1985

Eleanor G. Gaatman

Specification

Date received in lab.

Tests	Found	Requirements
Material	Soil	
Appearance	MEDIUM BROWN IN COLOR CONTAINS STONES OF VARIOUS SIZES	
Natural Water Content, %	4.7	
Sieve Analysis, By weight passing, % Sieve Size:		
1 inch	85.2	
3/8 inch	71.6	
No. 4	61.8	
No. 8	55.1	
No. 40	27.5	
No. 200	6.1	
Plastic Limits, %	NONPLASTIC	Paul Kramer

GRAIN SIZE ANALYSIS

BOULDERS	COBBLES	GRAVEL			SAND			SILTS & CLAYS IDENTIFIED BY PLASTICITY
		COARSE	MEDIUM	FINE	COARSE	MEDIUM	FINE	
228	76.2	25.4		9.52	2.0	0.59	0.25	0.074 MM
9"	3"	1"		3/8"	No.10	No.30	No.60	No.200 Sieve

DESCRIPTION

LOCATION
REMARKS

Picked
Sample No. 2 Depth
Lab. No.
Date
LL ___ PL ___ PI ___
SG ___ By ___

(11)

DEPARTMENT OF GENERAL SERVICES
LABORATORY - 480 Canal St., NYC

Date 7/31/85

CITY OF NEW YORK
DEPARTMENT OF GENERAL SERVICES
Bureau of Laboratories

Lab. # BG-0886
Sample 3

Insp. # _____ Lab. # _____
Dept. General Services Sect. _____
Material Soil   Spec. # _____
Brand _____ Contract # _____
Contract _____
Point/Date of Sampling Old subway tunnel at Atlantic
Remarks Avenue & Court Street
Sampled by: ____Joseph____
              (Signature)

Insp. # _____

Contract # _____

INFORMATION

SAMPLE 3 WAS TAKEN FROM CORE LOCATION No. 6
79 - Date Rec'd by Lab. **JUL 23 1985**

Specification   Date received in lab.

Tests	Found	Requirements
Material	Soil	
Appearance	MEDIUM BROWN IN COLOR CONTAINS STONES OF VARIOUS SIZES	
Natural Water Content, %	3.2	
Sieve Analysis, By weight passing, % Sieve Size:		
1 inch	76.2	
3/8 inch	69.0	
No. 4	62.6	
No. 8	56.5	
No. 40	25.1	
No. 200	3.6	
Plastic Limits, %	NONPLASTIC (12)	Paul Kringer

Sheet #13 of 13 Sheets

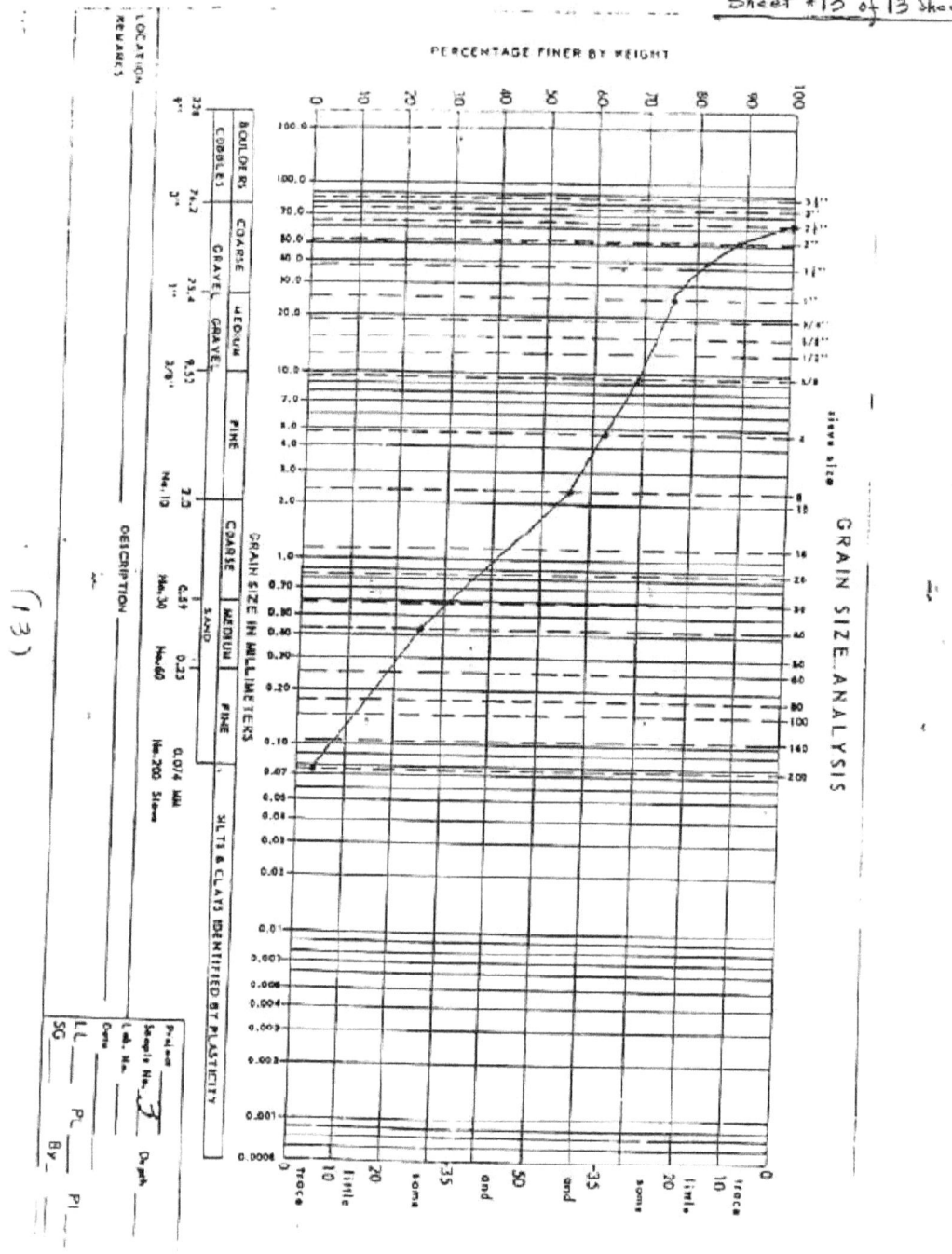

New York City Transit Authority

370 Jay Street, Brooklyn, New York, 11201 Phone (718) 330-

Members of the Board
Robert R. Kiley
Chairman
Lawrence R. Bailey
Daniel T. Scannell
Vice Chairman
Stephen Berger
Laura Blackburne
Stanley Brezenoff
Thomas Egan
Herbert J. Libert
John F. McAlevey
Ronay Menschel
Constantine Sidamon-Eristoff
Robert F. Wagner, Jr.
Robert T. Waldbauer
Alfred E. Werner

David L. Gunn
President

January 9, 1986

Mr. Joseph W. Ketas
Director, CEQR
Department of City Planning
Two Lafayette Street, Room 2400
New York, N.Y. 10007

Re: CEQR-85-003K
ATLANTIC AVE. TUNNEL PROPOSAL

Dear Mr. Ketas:

This letter is in response to your December 23, 1985 letter to C. L. Turin enclosing an engineering report submitted by the applicant of the subject project.

We have made a cursory review of the November 27, 1985 report by Singstad, Hurka & Associates submitted with your December 23 letter. We have no comments on this report except to note that on pg. 2 a live load of 500 P.S.F. was used in the design. The Transit Authority uses a live load that varies with depth. The attached Table 3 indicates the loading which, for a cover of 10 feet, is 0.6 K.S.F. The Consultant is experienced in the design of tunnels and can be relied on for a proper recommendation.

Very truly yours,

M. Oberter, P.E.
Division Engineer

320:NK:ss
123085 NK/L/ss
Attachment

Copy To: Mr. Robert Diamond, President
Brooklyn Histroical Railway Assoc.
599 E. 7th Street
Brooklyn, N.Y. 11218

Honorable Jack Lusk
Office of the Mayor
52 Chambers Street (Rm 108)
New York, N.Y. 10007

Table 3 gives the uniform live load, together with the corresponding dead load, for various covers of dry earth.

Table 4 gives the equivalent total load per sq. ft. which for various covers, spans and spacings of roof beams produces in the latter the same moment or shear as does the local concentration specified in Subsection (b) above, together with the corresponding dead load, for various covers of dry earth.

TABLE 3

SIDEWALK AND ROADWAY LOAD OVER SUBWAYS IN KIP PER SQ. FT.

Cover in Ft.	Dead Load (Cover Load)	Live Load		Total Load	
		Sidewalk	Roadway	Sidewalk	Roadway
2	0.2	0.6	1.3	0.8	(1.5)
3	0.3	0.6	1.2	0.9	(1.5)
4	0.4	0.6	1.1	1.0	(1.5)
5	0.5	0.6	1.0	1.1	1.5
6	0.6	0.6	0.9	1.2	1.5
7	0.7	0.6	0.8	1.3	1.5
8	0.8	0.6	0.7	1.4	1.5
9	0.9	0.6	0.6	1.5	1.5
10	1.0	0.6	0.6	1.6	1.6
11	1.1	0.6		1.7	
12	1.2	0.6		1.8	
13	1.3	0.6		1.9	
14	1.4	0.6		2.0	
15	1.5	0.5		2.0	
16	1.6	0.4		2.0	
17	1.7	0.3		2.0	
18	1.8	0.2		2.0	
19	1.9	0.1		2.0	
20	2.0	0.0		2.0	

For each additional ft. of cover, increase total load by 0.1 kip per sq. ft.

Values in brackets are minimum values and shall be compared to those given in Table 4.

For roofs below water or with depressed ceiling, increase load as specified in Sec. 2.

APPENDIX B

Table 1: AREAS WITH 1985-87 OZONE EXPECTED EXCEEDANCES GREATER THAN 1.0

EPA REGION	METROPOLITAN AREA (CMSA/MSA)	1985-87 DESIGN VALUE	AVG. EST. EXC	1987 2ND DAILY MAX 1-HR	Est. EXC
I	Boston, MA (CMSA)	0.14	2.2	0.14	4.3
I	Conn./Mass., CT-MA (Note #4)	0.17	5.8	0.17	11.6
I	*Hancock County, ME	0.13	1.3	0.12	1.1
I	*Kennebec County, ME	0.12	1.2	0.09	0
I	*Knox County, ME	0.15	4.4	0.13	6.5
I	*Lincoln County, ME	0.13	2.4	- NO DATA -	
I	New Bedford, MA	0.14	2.4	0.12	1.0
I	Portland, ME	0			
I	Portsmouth				
I	Providence, RI-MA (CMSA)	0.16	6.5	0.16	7.8
I	Worcester, MA	0.13	2.1	0.11	0
I	*York County, ME	0.15	4.2	0.14	4.9
II	Atlantic City, NJ	0.14	3.4	0.14	4.0
II	*Jefferson County, NY	0.13	4.7	0.13	4.7
II	New York, NY-NJ-CT (CMSA)	0.19	7.5	0.19	19.2
III	Allentown-Bethlehem, PA-NJ	0.13	1.4	0.13	3.2
III	Baltimore, MD	0.17	7.9	0.17	11.1
III	Huntington, WV-KY-OH	0.14	3.8	0.14	5.2
III	*Kent County, DE	0.13	1.8	0.15	3.2
III	Norfolk, VA	0.13	2.0	0.13	2.0
III	Parkersburg, WV-OH	0.13	1.5	0.15	3.5
III	Philadelphia, PA-NJ-DE (CMSA)	0.16	13.6	0.18	23.2
III	Pittsburgh, PA (CMSA)	0.13	1.7	0.14	4.1
III	Richmond, VA	0.13	1.3	0.14	3.0
III	Washington, DC-MD-VA	0.15	6.2	0.16	10.5
IV	Atlanta, GA	0.17	13.5	0.17	15.0
IV	Birmingham, AL	0.15	3.2	0.14	3.1
IV	Charlotte, NC-SC	0.13	3.0	0.14	4.0
IV	Jacksonville, FL	0.16	2.1	0.12	1.1
IV	Lexington, KY	0.13	1.6	0.11	1.1
IV	Louisville, KY	0.16	4.0	0.13	2.0
IV	Memphis, TN-AR-MS	0.13	2.0	0.13	2.0
IV	Miami-Hialeah, FL (CMSA)	0.15	2.1	0.15	3.1
IV	Montgomery, AL	0.14	2.2	0.14	4.3
IV	Nashville, TN	0.14	3.2	0.14	3.2
IV	Raleigh-Durham, NC	0.13	1.4	0.13	3.2
IV	Tampa, FL	0.13	2.1	0.16	4.2
V	Chicago, IL-IN-WI (CMSA)	0.17	7.4	0.18	12.8
V	Cincinnati, OH-KY-IN	0.14	1.6	0.15	2.1
V	Cleveland, OH	0.13	1.8	0.13	2.2
V	Detroit, MI (CMSA)	0.13	2.0	0.13	2.1
V	Grand Rapids, MI	0.13	1.3	0.14	3.0
V	Indianapolis, IN	0.13	1.3	0.12	1.1
V	*Kewaunee County, WI	0.13	1.9	0.14	5.8
V	Milwaukee, WI (& Sheboygan, WI)	0.17	3.7	0.20	12.9
V	Muskegon, MI	0.17	6.0	0.18	11.0

EPA REGION	METROPOLITAN AREA (CMSA/MSA)	1985-87 DESIGN VALUE	AVG. EST. EXC	1987 2nd Daily MAX 1-HR	EST. EXC
VI	Baton Rouge, LA	0.14	3.0	0.16	5.1
VI	Beaumont-Port Arthur, TX	0.13	2.1	0.13	3.2
VI	Dallas-Fort Worth, TX (CMSA)	0.16	6.1	0.14	.5.2
VI	El Paso, TX	0.16	9.0	0.17	11.1
VI	Houston, TX (CMSA)	0.20	19.1	0.18	20.8
VI	*Iberville Parish, LA	0.13	2.4	0.13	2.1
VI	Tulsa, OK	0.12	1.1	0.12	1
VII	St. Louis, MO-IL	0.16	5.4	0.17	8.0
VIII	Salt Lake City, UT	0.15	3.8	0.11	1.0
IX	Bakersfield, CA (Note #5)	0.16	35.1	0.16	47.6
IX	Fresno, CA	0.17	30.5	0.17	42.6
IX	*Kings County, CA	0.13	5.6	0.13	5.6
IX	Los Angeles, CA (CMSA)	0.35	143.5	0.32	141.2
IX	Modesto, CA	0.15	16.2	0.15	20.8
IX	Phoenix, AZ (Note #5)	0.14	2.4	0.11	0.
IX	Sacramento, CA (Note #5)	0.17	9.7	0.17	14.6
IX	San Diego, CA	0.18	12.5	0.18	26.8
IX	San Francisco, CA (CMSA)	0.14	3.4	0.15	4.1
IX	Santa Barbara, CA	0.14	1.7	0.13	3.4
IX	Stockton, CA (Note #5)	0.14	8.1	0.12	(inc.)
IX	Visalia, CA (Note #5)	0.15	11.9	0.15	21.6
X	Portland, OR-WA (CMSA)	0.15	1.8	0.11	1.2

* Not a Metropolitan Statistical Area

NOTES:

1. Metropolitan Statistical Areas are defined by the Office of Management and Budget, and include a central county and adjacent counties, if any, which interact with the urban area.

2. The air quality design value is the fourth highest monitored value with 3 complete years of data since the standard allows one exceedance for each year. This value may differ from the actual SIP control strategy value due to air quality modeling considerations such as the level of transported ozone.

3. The National Ambient Air Quality standard for ozone is 0.12 parts per million (ppm) daily maximum 1-hour average not to be exceeded more than once per year on average. The average estimated number of exceedances column shows the number of days the 0.12 ppm standard was exceeded on average at the site recording the highest design value after adjustment for incomplete, or missing days, during the three year period, 1985-87. The highest design value and the highest estimated exceedances for just 1987 are shown in the last two columns. These two values may be from two different monitoring sites.

4. Connecticut - Massachusetts includes Bristol, Hartford, Middletown, New Britain, New Haven, and New London, CT and Springfield, MA MSA's.

5. Incomplete data at this time, thus expected exceedance estimate is preliminary, however the air quality status with respect to the standard will not change.

Table 3

Areas With Two or More Exceedances of the Carbon Monoxide NAAQS, 1986-87

EPA REGION	METROPOLITAN STATISTICAL AREA (MSA)	1986 2nd MAX 8-HR (ppm)	# EXC	1987 2nd MAX 8-HR (ppm)	# EXC
I	Boston, MA	9.7	2	7.1	0
I	Hartford, CT	10.9	3	11.4	7
I	Manchester, NH	10.6	6	10.3	5
I	Nashua, NH	10.3	3	9.1	1
I	Springfield, MA	9.7	2	8.9	1
II	Bergen-Passaic, NJ	10.0	2	8.3	0
II	Jersey City, NJ	9.7	2	8.0	0
II	Nassau-Suffolk, NY	8.9	1	9.9	4
II	New York, NY (Note #4)	15.1	40	19.6	86
II	Newark, NJ	11.7	3	8.9	1
II	Syracuse, NY	11.3	6	9.6	2
III	Baltimore, MD	12.3	5	9.2	1
III	Pittsburgh, PA	10.4	2	8.8	1
III	Washington, DC-MD-VA	8.6	0	11.4	2
IV	Memphis, TN-AR-MS	11.9	2	10.5	2
IV	Nashville, TN	10.2	3	8.5	1
IV	Raleigh-Durham, NC	13.9	20	9.7	2
V	Cleveland, OH (Note #5)	10.1	2	6.6	0
V	Detroit, MI (Warren, MI)	11.9	2	9.4	1
V	Duluth, MN	9.6	2	8.5	1
V	Minneapolis-St. Paul, MN-WI	9.7	5	13.3	5
V	Steubenville-Weirton, OH-WV	9.1	0	19.1	24
VI	Albuquerque, NM	12.7	15	16.3	14
VI	El Paso, TX	12.1	10	15.4	11
VI	Houston, TX	9.8	2	8.3	0
VI	Oklahoma City, OK	10.7	3	11.4	4
VII	Lincoln, NE	9.9	3	7.2	0
VII	Springfield, MO	9.5	2	7.5	1
VII	St. Louis, MO-IL	8.6	0	10.5	2
VII	Wichita, KS	9.6	2	9.0	0
VIII	Denver, CO	25.8	33	15.6	24
VIII	Fort Collins, CO	12.4	6	12.8	5
VIII	*Great Falls, MT	9.3	1	11.0	3
VIII	Greeley, CO	11.6	4	10.5	3
VIII	*Missoula, MT	8.9	1	10.6	4
VIII	Provo-Orem, UT	14.4	24	13.3	20
VIII	Salt Lake City-Ogden, UT	11.6	9	9.8	2

Table 3 - Continued

Areas With Two or More Exceedances of the Carbon Monoxide NAAQS During 1986-87

EPA REGION	METROPOLITAN STATISTICAL AREA (MSA)	1986 2nd MAX 8-HR (ppm)	# EXC	1987 2nd MAX 8-HR (ppm)	# EXC
IX	Anaheim-Santa Ana, CA	10.0	3	9.7	2
IX	Chico, CA	10.1	2	7.9	0
IX	Fresno, CA	15.6	8	9.9	2
IX	Las Vegas, NV	15.9	27	16.0	20
IX	Los Angeles-Long Beach, CA	18.1	54	16.9	40
IX	Modesto, CA	11.1	4	7.1	0
IX	Phoenix, AZ (Note #6)	16.0	78	11.2	11
IX	Reno, NV (Sparks, NV)	13.4	23	8.9	1
IX	Sacramento, CA (& S.Lake Tahoe)	12.5	11	12.3	9
IX	San Francisco, CA	10.4	2	8.5	1
IX	San Jose, CA	10.6	4	7.1	0
IX	Vallejo-Fairfield-Napa, CA	10.1	4	8.3	0
X	Anchorage, AK	11.7	5	11.5	4
X	Boise City, ID (Note #7)	9.7	3	8.3	0
X	*Fairbanks, AK	14.8	32	13.9	15
X	*Grants Pass, OR	10.2	2	9.7	4
X	Medford, OR	12.6	18	9.5	3
X	Seattle, WA (& Bellevue)	11.9	6	10.0	4
X	Spokane, WA	16.0	44	19.0	66
X	Tacoma, WA	12.2	6	14.8	9
X	Vancouver, WA	- NO DATA -		9.8	2
X	Yakima, WA	11.0	3	11.0	4

* Not a Metropolitan Statistical Area

NOTES:
1. Metropolitan Statistical Areas are defined by the Office of Management and Budget, and include a central county and adjacent counties, if any, which interact with the urban area.

2. The National Ambient Air Quality Standard for carbon monoxide is 9 ppm 8-hour nonoverlapping average not to be exceeded more than once per year. The rounding convention in the standard specifies that values of 9.5 ppm, or greater, are counted as exceeding the level of the standard.

3. The exceedances of the carbon monoxide standard listed in the table are from the same site which recorded the highest second maximum 8-hour concentration in that year.

4. The exceedance count for New York in 1986 is from the second highest site because the site with the highest value had incomplete data for that year.

5. The monitoring site recording the violation of the standard in 1986 was discontinued early in 1986.

6. The site shown for 1986 was discontinued at the beginning of 1986. The site shown for 1987 recorded a 1986 2nd max 8-hour value of 13.5 ppm with 24 exceedances.

7. The two exceedances in 1987 which resulted from a building fire adjacent to the monitoring site met the criteria for exceptional events and are not shown.

APPENDIX C

APPENDIX "C"

Calculations:

Given:

As per Scenario "B" of the MetroTech space allocation (Table II-2 of the MetroTech EIS), the space allocations are as follows:

- 1,042,000 sq. ft. of Academic Space

- 3,000,000 sq. ft. of Commercial (Office/R&D) Space

- 189,000 sq. ft. of Office (Retail) Space

As per Table IV-2 of the MetroTech EIS, the person trip Generation Rate Per Day is as follows:

- Academic: 26.6 trips/1,000 sq. ft.

- Commercial (Office): 17.3 trips/1,000 sq. ft.

- Office (Retail): 22.0 trips/1,000 sq. ft.

As per Table IV-6 of the MetroTech EIS, the Geographic Distribution of Person-Trips to MetroTech from Long Island and New Jersey are:

-9.1% from Long Island, with 36.7% of these person trips by car

-2.6% from New Jersey, with 27.9% of these person trips by car

Therefore:

Total number of daily trips to MetroTech (by Space Allocations and Daily Trip Generation Rates) :

-Academic :   26.6/ 1,000 sq. ft. x 1,042,000 sq. ft.
                = 27,717.2 person trips

-Commercial (Office/R&D) :   17.3 trips/1,000 sq. ft. x
                3,000,000 sq. ft.
                = 51,900.0 person trips

—Commercial (Retail) :   22.0 trips/1,000 sq. ft. x

189,000 sq. ft.

= 4,158 person trips

- - - - - - - - - - - - - - - -

Total Person Trips to
MetroTech/Day:
83,775.2

Daily Trips to MetroTech from Long Island by Car:

−83,775.2 trips x 0.091 x 0.367 = $\underline{2797.8}$
 person trips per day from Long Island by car

Daily trip to MetroTech from New Jersey by Car:

−83,775.2 x 0.026 x 0.279 = $\underline{607.7}$ per trips
 per day from New Jersey by car

TOTAL : 3,405.5 person trips per day by car from
         Long Island and New Jersey

3,405.5 trips/day x 250 days = 851,375
person trips per year by car from Long Island
and New Jersey

TABLE IV-2

TRIP GENERATION RATE SUMMARY

Land Use Category	Daily Person Trip Rate
Academic	
o Employees	6.8 trips/1000 sq. ft.
o Students	16.8 trips/1000 sq. ft.
o Visitors	3.0 trips/1000 sq. ft.
	26.6 trips/1000 sq. ft.
Commercial (Office)	
o Employees	14.3 trips/1000 sq. ft.
o Visitors	3.0 trips/1000 sq. ft.
	17.3 trips/1000 sq. ft.
Office/Retail	
o Employees	2.0 trips/1000 sq. ft.
o Patrons	20.0 trips/1000 sq. ft.
	22.0 trips/1000 sq. ft.

Source: Urbitran Associates

METROTECH
ENVIRONMENTAL IMPACT STATEMENT

TABLE II-2
METROTECH PROJECT COMPONENTS PHASE II

SCENARIO A

Site [1]	Office	Academic	Comm'l/ R&D	Retail	Total Phase II
Site D (Expansion)	0	100,000	0	0	100,000
Site E	0	0	300,000	20,000	320,000
Site F	0	0	803,000	50,000	853,000
Site G	0	0	707,000	35,000	742,000
Site H	0	0	268,000	27,000	295,000
TOTAL	0	100,000	2,078,000	132,000	2,310,000

PROJECT TOTALS — SCENARIO A

Office	Academic	Comm'l/ R&D	Retail	Total Phases I&II
822,000	642,000	2,578,000	189,000	4,231,000

SCENARIO B

Site [1]	Office	Academic	Comm'l/ R&D	Retail	Total Phase II
Site D	0	100,000	0	0	100,000
Site E	0	0	300,000	20,000	320,000
Site F	0	0	803,000	50,000	853,000
Site G	0	291,000	416,000	35,000	742,000
Site H	0	109,000	159,000	27,000	295,000
TOTAL	0	500,000	1,678,000	132,000	2,310,000

PROJECT TOTALS — SCENARIO B

Office	Academic	Comm'l/ R&D	Retail	Total Phases I&II
822,000	1,042,000	2,178,000	189,000	4,231,000

[1] Site locations are shown on Exhibit II-5.
[2] Existing Polytechnic University space.

TABLE IV-6
GEOGRAPHIC DISTRIBUTION OF PEAK HOUR PERSON-TRIPS AND VEHICLE TRIPS

Origin/Destination	Person-Trip % Distribution	Peak Period % Auto	Vehicle-Trip % Distribution
Bronx	5.4	22.6	6.0
Brooklyn	55.3	16.1	41.0
Manhattan	5.7	9.5	2.0
Queens	15.9	30.5	22.0
Staten Island	4.0	45.1	8.0
Westchester	1.5	37.4	3.0
Long Island	9.1	36.7	15.0
New Jersey	2.6	27.9	3.0
Total	100.0	NA	100.0

Source: Urbitran Associates

APPENDIX D

Traffic Intersections Impacted
by Metrotech and ATURA

TABLE IV-13
SIGNIFICANT PROJECT IMPACTS AT SIGNALIZED INTERSECTIONS - 1993

Link	Movement	Time	No-Build V/C	Build V/C
Adams/Tillary	SBL	AM	1.18	1.26
Flatbush/Tillary	NBR/T	AM	1.07	1.08
	NBL	AM	1.07	1.08
	WBR	AM	1.51	1.52
	WBT	AM	1.40	1.42
	WBL	AM	1.32	1.92
	EBR/T	AM	1.24	1.28
	EBL	AM	1.00	1.20
	NBR/T	MD	1.32	1.41
	NBL/T	MD	1.32	1.41
	WBR	MD	1.73	2.00
	EBR/T	MD	1.16	1.21
	NBR/T	PM	1.02	1.08
	NBL	PM	1.02	1.08
	WBL	PM	1.51	1.57
	EBR/T	PM	1.37	1.41
	EBL	PM	1.16	1.58
Jay/Johnson	NBall	AM	.74	1.31
	NBall	MD	.52	1.05
	NBall	PM	.85	1.66
Jay/Tillary	NBR/T	AM	.85	.98
	NBL	AM	.85	.98
	NBR/T	PM	1.16	1.52
	NBL	PM	1.16	1.52
Flatbush/Myrtle	EBall	PM	.89	1.25
Flatbush/Willoughby	SBall	PM	1.13	1.20
Gold/Tillary	WBall	AM	.89	1.01
	WBR/T	AM	1.29	1.46
	WBR	PM	1.11	1.13
	WBT	PM	1.19	1.21
Bridge/Tech	EBall	AM	.97	1.35
	EBall	MD	.53	.88
	EBall	PM	.97	1.03
Bridge/Willoughby	WBall	AM	.87	1.54
	WBall	MD	.59	.93
	WBall	PM	.68	1.87

continued

TABLE IV-13
SIGNIFICANT PROJECT IMPACTS
AT SIGNALIZED INTERSECTIONS - 1993
(continued)

Link	Movement	Time	No-Build V/C	Build V/C
Jay/Willoughby	NBall	AM	.75	1.17
	WBall	AM	1.37	2.14
	WBall	MD	1.27	2.11
	WBall	PM	1.35	3.93
Jay/Fulton	NBall	AM	.96	1.56
	NBall	PM	.71	.85
	SBall	PM	1.34	1.54
Flatbush/DeKalb	SBall	PM	1.03	1.10
Flatbush/Fulton	WBall	AM	1.19	1.21
	SBall	PM	1.10	1.15
Flatbush/Atlantic	WBall	AM	1.37	1.49
	WBall	MD	1.44	1.49
	EBall	AM	1.36	1.37
	EBall	MD	1.19	1.25
	WBall	PM	1.25	1.27
	EBall	PM	1.20	1.33
Schmermerhorn/3rd	EBall	PM	.99	1.10

1991 Build Volume to Capacity Ratios (8–9 AM)
Figure IIF-40

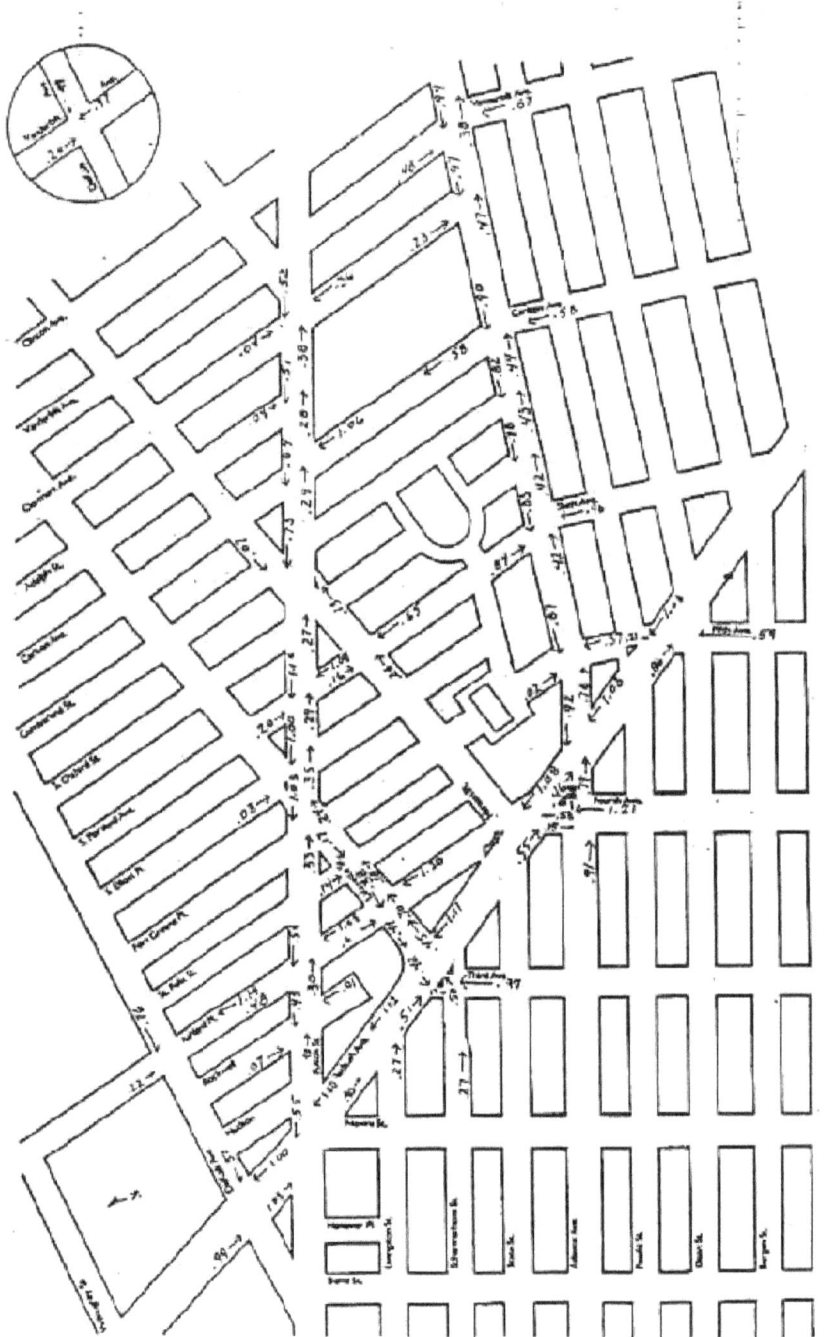

Note: V/C Ratios above 1.00 reflect demand volumes that cannot be fully processed at that street's theoretical capacity.

Revised from DEIS.

Atlantic Terminal/Brooklyn Center

1991 Build Volume to Capacity Ratios (12–1 PM)
Figure IIF-41

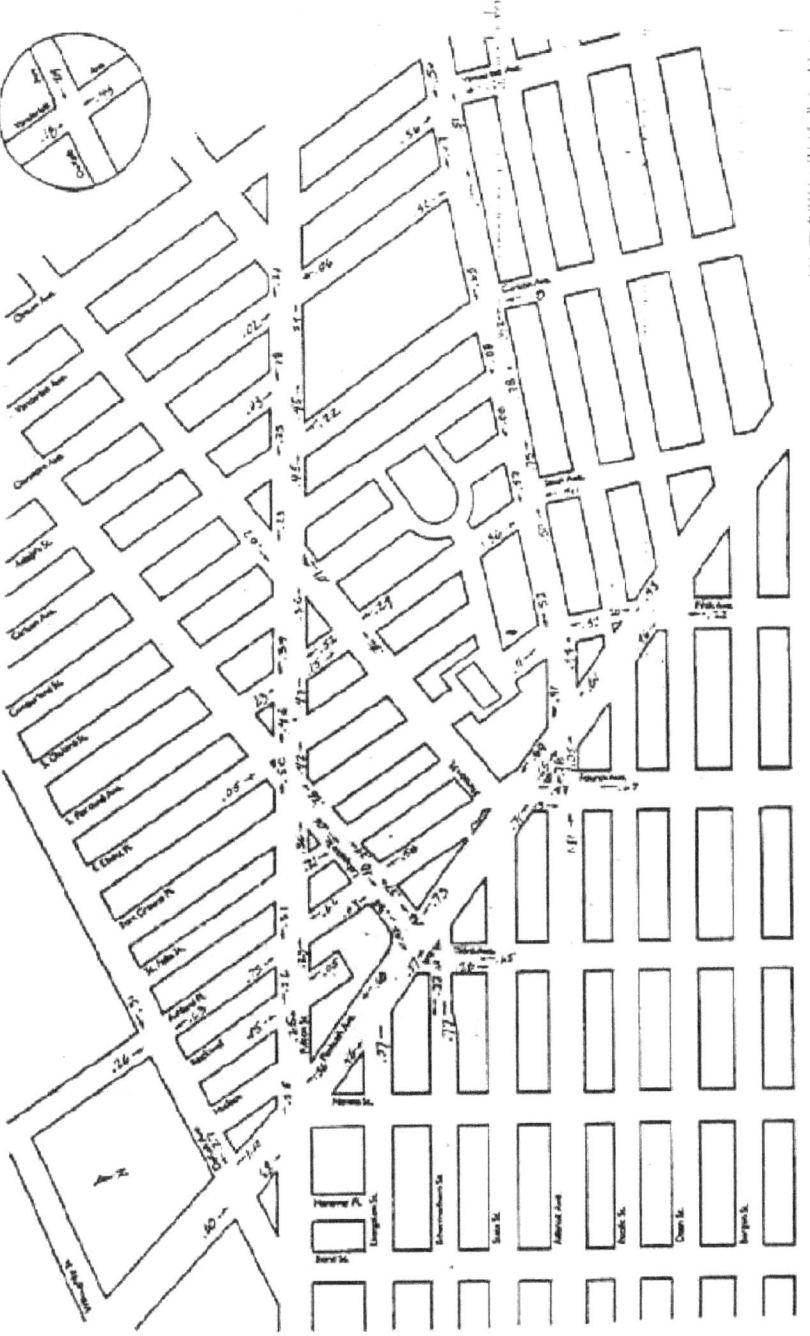

Note: V/C Ratios above 1.00 reflect demand volumes that cannot be fully processed at that street's theoretical capacity.

Revised from DEIS.

Atlantic Terminal/Brooklyn Center

1991 Build Volume to Capacity Ratios (5–6 PM)
Figure IIF-42

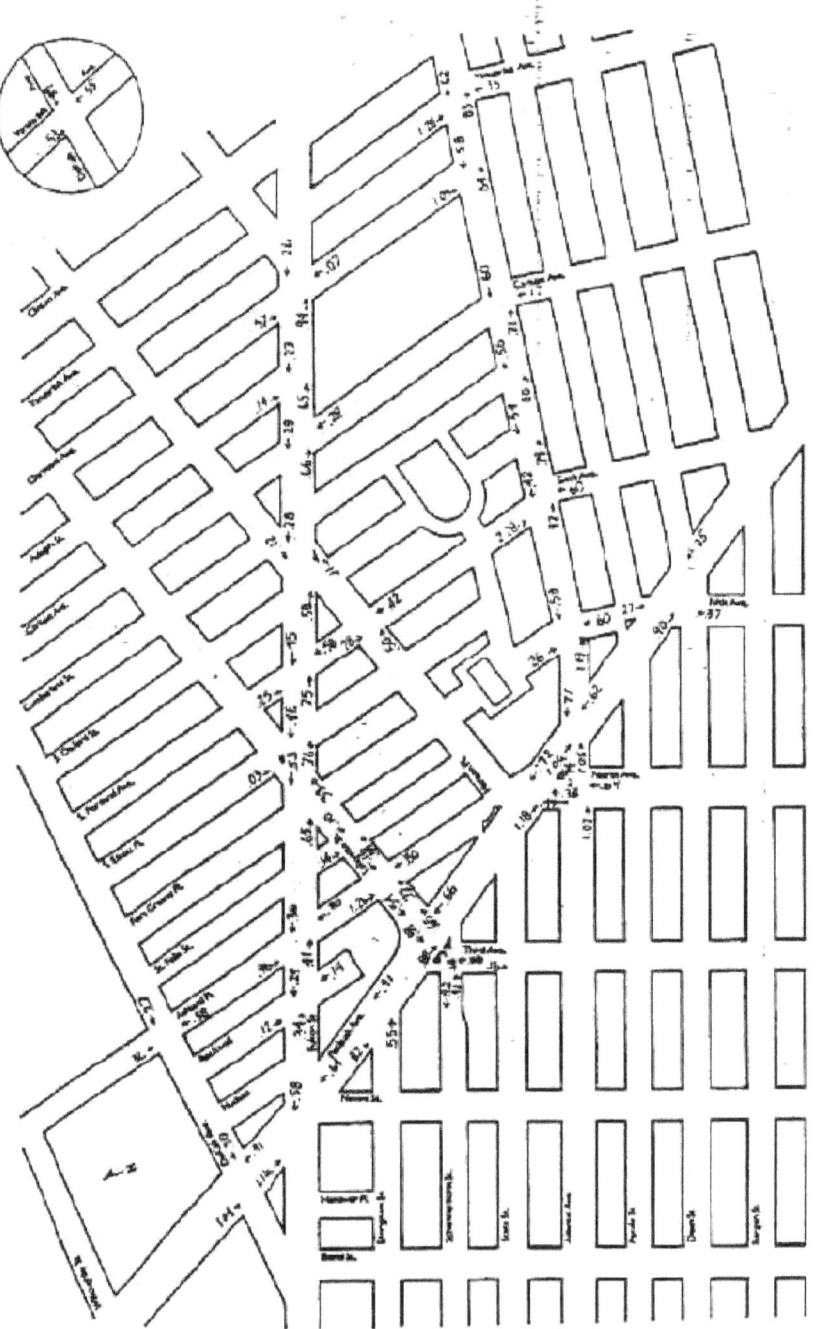

Note: V/C Ratios above 1.00 reflect demand volumes that cannot be fully processed at that street's theoretical capacity.

Revised from DEIS.

Atlantic Terminal/Brooklyn Center

APPENDIX F

Metrotech Subway Impacts

The full development of Metrotech would increase the number of persons entering the Jay Street IND station through the Myrtle Avenue turnstiles by 204 passengers in the 5-minute PM peak or approximately 40 passengers per minute. Substantial increases in existing passenger flows due to Metrotech would increase conflicts between queueing passengers at the token booth and the flow of passengers heading to and from the turnstiles. Measures would be required to provide additional queuing space and to allow the queue to locate in an area that would not restrict access to the turnstile.

Lawrence Street Station

Under 1989 Build conditions, the level of service at the Lawrence Street Station would decrease from A to C during both the AM and PM peaks, but would still be within acceptable standards. Under 1993 Build conditions, the level of service would significantly decrease from A to D at the Willoughby/Lawrence Street stairway and from A to B at the Willoughby/Bridge Street stairway during both periods. The impacts at the Willoughby/Lawrence Street stairway, resulting in v/c ratios of 1.2 in both peaks vs. 0.4 - 0.5 in the No Build condition, would represent a significant impact of the project. Widening of the stairway to mitigate this impact would not be feasible due to space limitations on the sidewalk and within the station's mezzanine. No other practicable mitigation measures are available, therefore, this impact would remain unmitigated.

Borough Hall Station (4,5)

Under 1989 and 1993 Build conditions, the additional traffic due to Metrotech would have no significant impact on the projected level of service for any of the station's stairway.

Borough Hall Station (2,3)

Under 1989 and 1993 Build conditions, the increase in the number of projected generated passengers using this station would not change operating conditions.

(2) Subway Line Capacity Impacts

Under 1989 Build conditions, the additional project loads by line range from a low of 53 trips for the outbound afternoon IRT No. 2 and 3 service to a high of 754 morning peak hour trips on the BMT M and RR inbound service. Comparing existing and future passenger loads with planning capacity levels, the additional passengers due to Phase I of Metrotech would worsen scheduled capacity shortfalls on the IRT 4 and 5 lines Outbound in the AM, the IRT 3 and 4 lines Inbound in the AM and Outbound in the PM, and the IND "F" line Outbound in the PM.

Metrotech
Environmental Impact Statement

IV. IMPACTS OF PROPOSED PROJECT

could be reduced to minimize the impact on total route cycle time. Bus operating costs would increase slightly due to the lengthening of bus routes, which would only affect non-labor costs. However, these cost changes would be relatively minor. For example, using an approximate value of $1 per mile for variable non-labor costs, (fuel, tires, etc.), a 600-foot increase in route length on the B54 route would add approximately 11 cents per run. The equivalent figure for the B61 route (approximately 4,000 feet of additional route length) would be 73 cents.

(3) **1993 No-Build Conditions - Public Transportation**

(a) **Subway Station and Line Capacity Impacts**

The 1993 No-Build volumes Metrotech area subway stations encompass the 1989 No-Build volumes, Phase II of the Atlantic Terminal project and a 0.5 percent annual growth factor for the background subway traffic from 1989 to 1993. From these numbers, it is projected that total growth in passenger movement for the impacted stations would amount to less than 20 persons per hour. As a result, all levels of service and line capacity conditions would remain the same as those noted for 1989 No-Build volumes. Table IV-22 outlines peak 5-minute utilization levels at all affected stairways.

(b) **Bus Operations**

Bus operating conditions and load factors on Metrotech-related routes under 1993 No-Build conditions will be essentially the same as under 1989 No-Build conditions.

(4) **1993 Build Condition - Public Transportation**

(a) **Subway System**

The full development of Metrotech would increase project-generated subway ridership during the morning peak hour by 4,616 (including 1,740 from Phase I). Afternoon peak period volumes would increase to 4,070 (including 1,430 from Phase I). The number of peak hour subway passengers added by Phases I and II of Metrotech would be as follows:

Metrotech
Environmental Impact Statement

IV. IMPACTS OF PROPOSED PROJECT

Metrotech Project
Site Generated Subway Trips

Period	Inbound			Outbound		
	Phase I	Phase II	Total	Phase I	Phase II	Total
8:00-9:00 AM	1,620	2,687	4,307	120	189	309
5:00-6:00 PM	169	401	570	1,324	2,239	3,563

In order to estimate the impact of these trips on the subway system, it was necessary to allocate them according to origin, subway station and subway line. Tables IV-23 and IV-24 show the projected AM and PM peak hour arriving and departing subway trips, respectively. Trips are shown by origin, subway line and subway station. The process used to allocate these subway trips was the same as that followed for allocating subway trips under the 1989 Build conditions.

Trips from the project site in the morning and to the site in the afternoon, in the non-peak hour direction, were also allocated to the four nearest stations, Jay Street, Lawrence Street, Borough Hall (4,5) and Borough Hall (2,3). These trips were, in turn, allocated to the lines serving those stations.

The next step was to allocate all trips to individual station stairways and escalators and to assess the resulting level of service. The allocation of the trips to stairs and escalators is shown in Table IV-25. The following sections describe the level of service analyses for each of the effected stations in the study area.

Jay Street Station - Most of the passengers generated by Metrotech are assumed to use staircase No. 5 at Jay Street and Myrtle Avenue (see Exhibit III-11), which is closest to the site. Trips to the Duffield Building have been routed equally through that stairway and stairway No. 6 at Jay and Willoughby Streets. The Myrtle Avenue staircase would have LOS F and D in the AM and PM, respectively, under the 1993 No-Build condition. The Jay and Willoughby Street exit would be at LOS C in the AM and D in the PM under the 1993 No-Build scenario. The v/c values for the No Build are shown in Table IV-22.

APPENDIX G

time saving. Lastly, some 5 percent previously used modes other than the subway, and may have been diverted to it because of the new convenience."

Another instance of gaining amenity at the expense of money and travel time could be observed on express buses in New York City. In the early 1970s, a segment of the City's transit riders, with incomes 30 percent above the average for their area of residence, spent 60 percent more money and 2 percent more time for an express bus ride compared to the mode previously used, which in more than half the cases was the subway. The major gains were an *assured seat* and air conditioning, though a minority also had fewer transfers and a shorter walk.[10] Because express buses in New York City were introduced specifically to serve areas poorly served by rail transit, we have here a case where the bus offered comfort and amenity superior to rail.

The more usual evidence is that rail has greater attractive power than the bus, particularly for diverting auto users, if travel time and fare are similar. Thus, in 1960 the number of journeys to work by public transportation was on the average 30.5 percent higher between parts of the New York Region which were connected by subway or rail compared to those that were not, if development density as well as travel time and the cost differences were kept constant.[11] Similarly in 1964, the number of trips for all purposes made during the peak three hours by public transportation across the Hudson River was 27.9 percent higher between zones that had rail service. This higher use was entirely due to greater diversion from the auto, which cannot be explained by differences in development density, travel time, or travel cost in corridors with rail service.[12] The next chapter will show that the presence of a commuter rail, and particularly a rapid transit line, does suppress auto ownership in its immediate vicinity, compared to the surrounding area which is in varying degrees served by bus. In the case of commuter rail, the greater passenger attractiveness can also be attributed to the more liberal space standards and the superior ride quality. One can speculate that less tangible factors, such as a permanent physical identity and system connectivity, also favor rail over bus.

Nationwide trends in transit ridership are sometimes cited in support of the greater attractiveness of rail: between 1955 and 1970 bus ridership per round-trip route mile in the United States declined by 39 percent, while rapid transit ridership per track mile declined only 8 percent, and ridership per surface track mile actually increased by 27 percent.[13] This evidence unfortunately is weak, because there is no way of separating the relative differences in travel cost and travel time, as well as other extraneous factors, and thereby isolating the attractiveness of rail as such. Thus rapid transit inherently offers greater time savings than the average bus, and also serves high density Central Business Districts where the auto is least competitive. Moreover, after 1972, following 25 years of continuous decline, nationwide bus ridership picked up because of extensive subsidy and service improvement programs, while rapid transit continued to drop, largely because of precipitous employment losses in New York City, where over 70 percent of the nation's rapid transit ridership is concentrated. As for light rail, the increase merely shows that streetcar trackage was abandoned at a faster pace than passengers declined. Clearly, if lightly used lines are abandoned the average ridership per mile will increase.

RESPONSE TO NEW FACILITIES

Exhibit 1.2 presents some evidence of ridership increase which new transit facilities were able to achieve, along with indications of how much of the new ridership repre-

Public Transportation & LAND USE POLICY

Boris S. Pushkarev and Jeffrey M. Zupan

A REGIONAL PLAN ASSOCIATION BOOK